Danger Close: A True Story of a Forward Observer in Vietnam

Danger Close:
A True Story of a Forward Observer in Vietnam

Patrick John McNulty, III

Deeds Publishing | Athens, GA

Published by Deeds Publishing in Athens, GA
www.deedspublishing.com

Printed in The United States of America

Cover and interior design by Deeds Publishing

ISBN 978-1-951921-72-9

Books are available in quantity for promotional or premium use. For information, email info@deedspublishing.com.

First Edition, 2025 10 9 8 7 6 5 4 3 2 1

I dedicate this book to my parents. My father taught me the importance of hard work, perseverance, honesty, and the initiative to learn. My mother, who taught me empathy for others, an attitude of gratitude, always looking on the bright side, and developing my creativity.

Contents

Introduction

My father, Patrick "Pat" McNulty Jr., was a soldier who served in the Vietnam War as a forward observer and later became a battery commander and battalion adjutant at Fort Sill, Oklahoma. These words are a legacy for him. My mission is to paint a picture so that future generations can see the sacrifices he made for his country.

My father began to be open about his war experiences only after many years had passed. It was only in the past ten years that my father began to share his story with his family.

As I put this book together, I have learned that soldiers do not usually talk about the terrors of war, often because they believe the average person cannot comprehend them. Many veterans still feel the effects of war years later because of post-traumatic stress disorder (PTSD) and would feel like it was just yesterday. I want to recognize all the veterans who have carried their stories in silence to the grave. I also want to thank all the veterans who have put their lives on the line for their country.

My father felt this book did not need to be written, but it was a story that I felt needed to be told.

Sgt. Byron Kinnan's Testimonial

"There were never any easy days for an infantryman in the jungles of Vietnam; each man relied on the man next to him to survive. When 'shit hit the fan,' there was the confusion and fear that death was right around the corner felt by all those men. Yet it was the artillery forward observer (FO) that had to be cool and face death head-on—to relay the correct coordinates and to walk among those artillery explosions closest to the soldiers to protect their lives!

"To be precise and accurate, even while the battle raged around him, was the test of any FO. Patrick "Pat" McNulty passed that test many times and saved many men from coming home draped in our flag. To paraphrase the poem, "IF" by Rudyard Kipling—"If you can keep your head while those all about you are losing theirs, you'll be a man, my son." The test of any good soldier was the number of good men they helped bring home safely."

— ***Byron Kinnan, Delta dog***
3rd Battalion, 8th Infantry

CHAPTER ONE

Surrounded in the Jungle: Lessons in Leadership & Survival at Hill 800

From the stories my father, LT Pat McNulty, who served as a forward observer for a company of infantryman in Vietnam, told. He realized a day out in the field can be full of surprises, but not in a good way. An example would be this one very distressing instance when the North Vietnam Army (NVA), otherwise known as Charlie, made a sneak attack on his unit. The NVA crept out from the underground tunnels of the mountain near where the Americans were patrolling. The Americans were caught off guard, as they did not know of the NVA's presence beneath the mountain. The NVA creeped out from below, surrounded them in the mountain's triple canopy jungle, and attacked them while barely even seen, hidden in the trees.

McNulty was with Bravo Company, 3rd Battalion, 12th Infantry at that time. McNulty described Bravo Company, 3rd Battalion, 12th Infantry (3/12 Infantry) as a good unit with good leadership and good men, but just one problem is that they saw more than their share of combat. Initially, he was an FO for Delta Company, 3rd Battalion, 8th Infantry (3-8 Infantry). The 3-8 Infantry had more than enough FOs with them, but Bravo Company, 3rd Battalion, 12th Infantry, on the other hand, needed one. So McNulty was shifted over to the 3/12 Infantry.

January 19th, 1968, Bravo Company of the 3rd Battalion, 12th Infantry Regiment moved off of a Fire Support Base (FSB) north of Ben Het, Kontum Province, Republic of South Vietnam. The mission was to advance westward towards a hilltop known as "The Peanut", and contact and engage enemy forces threatening the FSB and Highway 512 to the south.

So that day started with that little walk in the jungle, as usual. To their relief, without the burden of having to carry a heavy rucksack that time. They just had with them their web gear, harness, canteens, ammo pouches, rifle, extra 20 mags of ammo, and some C-rats. After walking about four clicks (kilometers), the infantry point men (the leading soldier advancing through hostile territory) spotted a couple of North Vietnamese Army (NVA) trail watchers. The trail watchers sat at the cross point of two trails. Those trails were in the NVAs backyard, and they were at home.

Those trails were critically important to the NVA unit, and the trail watchers had an important job., They were tasked with directing other NVA troops bringing in food, ammunition, medical supplies, etc.. The infantrymen walking point halted and radioed their commander, asking for guidance.

The commanding officer (CO), Capt. Robert Morton and McNulty responded in unison, "Shoot them." McNulty then told the captain he was bringing in artillery. "I've been in this situation before, and it can turn to shit really fast," he said to Morton.

The infantrymen paused while artillery was being called in. Some were able to grab a quick bite to eat, and rest their weary legs and backs.

McNulty thought of calling over SP4 Timothy Nafe, his recon sergeant by title. Nafe was around 20 or 21 years young and once told McNulty that he could not understand why they were fighting in the Vietnam War.

McNulty had tried to explain to him the Domino Theory. Even then, he still did not get it—"Smart man," McNulty remarked as he told his story. McNulty called him over to give Nafe some training on how to call in artillery.

McNulty had not adjusted fire since he returned to 'Nam after he had been wounded in November of 1967 and hospitalized in Japan. Getting wounded made McNulty realize that he was not as immortal as he thought.

McNulty asked Nafe if he had adjusted fire before. He answered, "No." All the more urgent to get some rudimentary cross-training. McNulty then told him, "Today is your lucky day." He then proceeded to teach Nafe how to do it. (There were men in McNulty's previous FO party present that time that had observed him calling in fire before, but it had not crossed his mind to ask them to do it.)

Since it was Nafe who would be directing the artillery, he would be the one to talk directly to the Fire Direction Center (FDC) back at the firebase. As someone without prior experience in adjusting fire, Nafe did not know where they were located on the map, which was expected.

McNulty showed him where they were located and taught him to translate map coordinates into code. They had map coordinates in code because had those numbers been written directly, if Charlie somehow got hold of their maps, it would be trouble for them. Charlie would quickly find out where their troops are located, and he could either send a welcoming party or direct some artillery of their own.

McNulty instructed Tim, "We bring in the first round 500 meters from our position. To make sure we haven't made a mistake in map reading." So what did Nafe have in mind for the first round? He said, "One high explosive (HE) round at XXXXX."

McNulty corrected him, "No. The first round is always a smoke round." Further explaining, "You'll never see it in this triple canopy jungle, but you'll hear it pop. Having smoke come in first is just a double backup that your map reading is correct. We would all be very upset with an HE round landing on us."

Nafe then called in smoke. Then came the pop sound.

"That's perfect. Remember that sound," McNulty remarked. Then he asked, "Now what?" Nafe responded, "HE."

"Correct. Tell them, repeat with HE (High Explosive)," McNulty told him. Nafe then said what McNulty instructed into the radio. It came in fine.

"Now what?" McNulty asked again. "Drop 200," he answered. McNulty corrected him, "We only drop in 100-meter increments to be double sure." So Nafe called in, "Drop 100." It came in fine.

"Now what?" McNulty asked again. "Drop 100," Nafe answered. "Good," McNulty remarked. Nafe called in again, "Drop 100." It came in fine. "Now what?" McNulty asked again.

"Drop 100," Nafe answered. McNulty corrected him. "No, at 300 meters, we only drop in 50-meter increments. We're at "danger close" fire at 300 meters."

The next thing they did was to make sure that all of the tubes of artillery from the firebase (FB) were all lined up correctly so that the projectiles they called in would land at the same point. The number of tubes used in a firebase would depend on the caliber of the tubes or how the Division decided to break up the supporting artillery to serve infantry needs. Sometimes it would be five, sometimes six. That allowed them to fire more tubes at a time instead of just one.

On this day, they had six. So McNulty explained to Tim, "We're about okay, but first and last, we want to make sure all six tubes are landing at the same point. To do that, call in, original tube (Howitzer) repeat fire. Then have a second tube also fire, with a two-second delay after the first. Then we'll listen to where they land."

Nafe spoke into the radio what McNulty said; then the artillery people fired as asked.

They noticed the differences in the impact. "Crunch... Crunch." "So, what's next?" McNulty asked again.

"Bring the second tube to the left, 50, and then add 50," Nafe answered.

"Good call," McNulty remarked, then told Tim, "Have them repeat fire, the same sequence, with a two-second delay."

Nafe called it in; the rounds came out. "Crunch... Crunch." Then McNulty instructed Tim, "Tell him that was perfect. Both tubes are hitting the same point. We will stop firing. We have artillery at a known spot, "Record as target. If anything happens, an exact point from which to adjust the Howitzers."

Satisfied with how things turned out, McNulty informed Captain Morton, "We're good. We've got artillery."

While they were firing explosives, little did they know that right below the ground they stood on was an NVA basecamp which meant that besides the trail watchers, there were plenty more NVA soldiers around. The basecamp tunnel system was up to 70 to 90 feet below them. The NVA were known to have basecamps in tunnels beneath mountains and hills. Inside their tunnels were field hospitals, rooms, beds, and all kinds of stuff set up, enough to support hundreds of NVA soldiers.

At that time, Bravo Company knew there were NVA in the area, but intelligence had not indicated that such a large basecamp and concentration of NVA soldiers was present. McNulty figured that area that the NVA came out of might have been home to them for the past two to 22 months. Although he was not sure for exactly how long, it was obvious that the NVA were accustomed to the area.

Bravo Company of the 3rd Battalion, 12th Infantry found themselves deep within the territory controlled by the North Vietnamese Army (NVA), who were extremely familiar with the area. Anticipating the arrival of American soldiers, the NVA surrounded them by digging out hidden pits (known as foxholes), setting up defensive positions, and placing snipers high up in the trees. The NVA was particularly skilled at operating in a triple canopy jungle—a dense forest that limits visibility, much like standing on a foggy seashore.

In a triple-canopy jungle, soldiers on the ground are surrounded by thick underbrush, vines up to 10 feet high, young saplings reaching for sunlight, and towering mature trees overhead. This dense vegetation blocks sunlight and makes it hard to see more than 15 feet above the ground.

An American soldier looking up might not spot a sniper hidden in the treetops due to the dense foliage.

However, snipers positioned in the trees could easily see through the lower layers of saplings, vines, and bushes. This situation is like being on a foggy seashore: you might not see a ship approaching, but someone on the ship can easily see the shore.

McNulty had no idea that the Viet Cong frequently operated from the treetops. His personal experience hadn't exposed him to this threat, and he hadn't spent enough time in the area to learn about it from other members of Bravo Company.

After a year in a unit, American soldiers were rotated out and returned to the United States. The Army would introduce about 30 new soldiers, who would train for one to two weeks alongside experienced troops. During this time, they would patrol together to learn the terrain and observe enemy position setups. It wasn't until decades after his service in Vietnam that McNulty discovered this was a common enemy tactic.

Captain Morton gave the order for his men to move out. Tim Nafe stood and began moving forward. Unfortunately, just as he took his very first step, he became the first man down, with the first bullet hitting him right through the heart. He was dead before hitting the ground. He was the first target, likely because he was seen using the radio, and identified as someone important who had been directing artillery. He had the radio, and he did all the talking. So he caught the attention of an NVA through the "fog," and the NVA targeted him.

Out in battle, anything that can identify someone as an essential person puts them at high risk of being an automatic prime target. Troops who carried a radio, a pistol, or a carbine instead of a rifle would often be targeted. Or if they were stupid enough to have officer insignia brass worn while they were out in the field, they would be putting themselves into trouble. Because it was apparent that they were either an officer, a non-commissioned officer (NCO), or someone in the command and control group.

Once the leadership and communications had been taken out first, the entire unit would become directionless and easy to rout.

Bullets began to fly everywhere. McNulty related that they were like a swarm of hornets kicking up the ground all around him. The radio was still with Tim's dead body. McNulty needed it, so he yelled to someone close by Tim's body, "Throw me the radio!"

The captain reported to the higher up about their contact, and with no idea how chaotic it had become, they questioned Captain Morton on how serious the situation was. Feeling agitated from the terror the unit was experiencing while being asked about it, Captain Morton replied to them—as clear as seconds ago, he remembered, "I'll tell you how serious it is. I'll put a dead man on the phone to tell you!" referring to Nafe.

Amidst the chaos, McNulty had to act quickly as casualties increased every second. He immediately moved Nafe's artillery closer to target the North Vietnamese

Army (NVA) attackers, aiming to stop more enemies from joining the fight. He tried to find the enemy by pinpointing where the heaviest gunfire was coming from. However, with bullets flying in all directions, this was extremely difficult. It felt like they were caught in the middle of a circle of shooters.

While adjusting the artillery fire from their base, McNulty received a call from another unit offering help with their larger cannons, known as 155mm howitzers. "Hello, I'm with the 155s; can I help you?" they asked. He repositioned the 155mm artillery to fire further away from his initial support, making it easier to tell their shots apart from the ones already being used. This extra firepower would help block more NVA reinforcements.

Another call came in offering assistance. "I'm with the 175 unit. Can we assist?" they asked. McNulty jokingly replied, "No, I'm kind of busy right now." Even though he was joking, he actually needed their help.

This was great news for him, as he was already managing fire from three different artillery batteries.

McNulty found himself facing a situation that wasn't covered in his training: how to coordinate fire from three different artillery batteries while moving and trying to escape a deadly trap. Usually, coordinating multiple artillery batteries is the job of higher-ranking officers, not a company-forward observer like him, who was typically a lieutenant.

In such cases, a coordinated plan called Time on

Target (TOT) is made, where all artillery groups fire at the same time on a specific target—not on a moving or changing one. He couldn't take out a notebook to sketch where the last shots had landed.

Although this situation wasn't part of his training, he had to adapt quickly. Meanwhile, another artillery unit with 8-inch cannons offered help, which he gratefully accepted. McNulty decided that in this changing situation, he could adjust the close-in support as needed.

He planned to use the larger artillery to cover and block fires, stop NVA reinforcements, and protect his unit's retreat back to the base. He gave each artillery group specific coordinates to create a straight line that, when viewed from above, looked like an upside-down "L." The long part of the "L" would cover the path back to the base, while the short part would prevent the NVA from attacking from the side.

As the situation changed and they moved closer to the base, the "L" was adjusted to stay near Bravo Company, 3rd Battalion, 12th Infantry.

McNulty provided the units with coordinates to form a straight line along their left side, covering about two sections on the map. He instructed, "Fire anywhere along that line and extend your range by 100

to 200 meters, depending on your cannon type." This allowed him to adjust a large area of fire at once, effectively turning the big cannons into a barrier. By then, McNulty was directing fire from four artillery units, which brought him a sense of relief. He could focus on directing the smaller 105mm and 8-inch cannons accurately within 25 yards while using the larger 155mm and 175mm units to block the enemy at a close range of 300 meters.

After-action reports initially showed that only a few reinforced companies were attacking Bravo Company, but to McNulty, it felt like the entire North Vietnamese Army was involved.

The NVA had only a short time to operate without being targeted by American artillery. Once the enemy was within 25 meters, using artillery was too risky because of the danger of hitting their own troops.

Therefore, the NVA tried to "hug" the American unit—getting as close as possible. Beyond 25 meters, they were highly vulnerable—assuming the forward observer was still alive.

Slowly, over the course of four long gruesome hours, Bravo Company withdrew to the relative safety of the firebase. There came ambush after ambush after ambush. It was sheer hell, as McNulty described. McNulty later found himself all alone. He had no group with him. Ninety-nine percent of the time, my father was alone in this battle. He was not entirely aware how he wound up all alone. He was too busy adjusting fire. As he was adjusting fire, he positioned himself at the very tail end of the withdrawal to not bring in the fire too close to his unit, what with the lethal range of those big calibers.

He would also occasionally stop to check his bearing

on the map. He was so preoccupied with trying to call in artillery, that he wound up alone until two GIs who did not have rifles approached him and asked if they could have his. Curious about their rifles, McNulty asked them where theirs were. They replied they dropped theirs because they were carrying back a wounded soldier who unfortunately died.

Since his focus was on the map and the radio the whole time, he had not used his rifle. As a considerate man, McNulty was willing to lend them his rifle. He just wondered how he would get by throughout the battle without it, so he also asked, "What could you give me?" In return, they offered him a grenade.

So, McNulty said, "Okay, but you guys swear you're going to stay with me and protect me?"

"Yes," they replied.

Unfortunately, the chaos of the battle continued, and within two minutes, McNulty found himself alone again.

At the moment, McNulty was alone, but on the bright side, he got reliable assistance from the artillery people. The troops at the firebase, headed by Captain Robert Barrett, did a great job supporting the company. The firebase had only been established two to three days before McNulty went out with Bravo Company.

As they got closer to the firebase, McNulty decided to stop the 105s and used mortars instead, because the 105s were better suited for longer distances.

As they were finally nearing the firebase, Charlie's

small arms fire subsided. McNulty was thankful that nothing was directed at him, but he did have one problem—the mortars—he noticed mortars were dropping in their area, as well as the area outside the withdrawing Bravo Company. McNulty started to have a bad feeling about the mortars since he had a bad experience with mortars.

This went back to November 7th at Hill 724, when artillery from their own firebase fell in the area (also called short rounds). He was worried that the American mortars were impacting in his area. So, he ordered the mortars to check fire to verify whether they were the cause of the rounds in their area. The American mortars stopped, but mortars kept falling. The mortars were, indeed, not short rounds from their firebase, but from Charlie.

Satisfied that his unit wasn't going to be hit with friendly fire, he called off the check fire and began directing the American mortars once again.

As he reached 100 meters within the firebase, he encountered a major looking sharp in a clean uniform with polished boots. The major was well-meaning and kind enough to offer McNulty a drink from his canteen. To McNulty's surprise, it was ice-cold scotch—he does not like scotch, unfortunately.

He did not even recall whether he swallowed any of it at all or not. At that point, he yearned to finally arrive at the firebase and collapse. It had been a relief for him when he came back to the firebase, but there was one last thing he had to do.

When he returned, he still had to turn over the artillery to the major. Someone asked him if he had done it already. So, he went out to the trail to turn it over.

The battle at Hill 800 left a harsh aftermath for their unit, with several casualties and a lot of equipment lost. As McNulty heard initially at the firebase, the casualties from the battle due to the fog of war were around 17 killed in action (KIA) and 35 to 40 wounded in action (WIA) out of around 75 men. After the battle, it was only Nafe's body that had been brought back. The battle they fought was intense. Niceties like bodies were of low priority when the casualties are high.

Only two radios—his and the captain's—got back to the firebase. Many weapons were lost, as they may have been jammed, or the soldiers may have dropped them to carry the wounded.

Fortunately, it was later discovered that the death toll from the battle for their unit was 7, not 17. That was 7 too many.

After the battle, every day, the infantry would search in and around Hill 800 to recover and bring back the other six KIAs. Also, their B52s carpet-bombed Charlie's area after the battle. That caused extremely high losses for Charlie.

However, those that made it inside the tunnels before they dropped the bombs were spared from the bombing.

So those that survived the carpet-bombing soon left the area. Their infantry also resumed searching for their dead men, even after they had carpet-bombed the area.

CHAPTER TWO

Generations of Honorable McNultys

Two generations of the McNultys before my father, Partick John McNulty Jr. were also soldiers. My great grandfather, James McNulty, served in an Irish military organization, which was a volunteer army called the Irish Volunteers (IV). My grandfather, Partick John McNulty Sr. served in the US Armed Forces in the Navy. Their military service and stories would be relevant to share and deserve to be recognized.

The McNultys are of Irish roots. James, originally from Ireland, migrated to America before serving in the Irish

Volunteers. James had an older brother named Patrick in America who left Ireland earlier than him.

Patrick worked as an apprentice in the brick trade. It was a year after Patrick left that James followed him to America. James was sixteen years old at that time. James was very quiet and rarely ever spoke unless needed. Just like his older brother, James also wanted to learn a trade. That is why he started an apprenticeship in brickwork.

Also, as he worked as an apprentice, he found himself working with John B. Kelly, a friend who had later become renowned as an Olympic medal winner in rowing, and for his business in building houses. Because Kelly knew of James' reputation as an excellent worker, he became the foreman on the Philadelphia Post Office at 30th street station.

While in America, James met the love of his life, Annie. As he found the woman he wanted to spend his life with, he took her with him back to Ireland, and there, they got married. Their wedding took place in Saint Michael's Chapel in Creeslough, which is no longer a chapel but a graveyard. They lived in a town in County Donegal called Creeslough, where he grew up.

It was around 1914 when James joined the Irish Volunteers (IV), also known as the Irish Volunteer Force (IVF) or Irish Volunteer Army (IVA). It was a militia founded by nationalists of Ireland in 1913 with the primary aim: "to secure and maintain the rights and liberties common to the whole people of Ireland." Because for a long time, the British had Ireland under their control. So, a group of nationalists came together to form the IV in an attempt to achieve freedom from British rule. The establishment of the IV is in connection to the British government's plans for the devolution of Ireland with the proposed Home Rule Bill would allow Ireland to have a separate government of its own.

When British Prime Minister William Ewart Gladstone first introduced that bill in 1896, a group of Irish Unionists, who were in favor of the union with Great Britain, was against it and formed a military organization called the Ulster Volunteers (UV) or Ulster Volunteer Force (UVF) in 1912. They were willing to fight in order to resist the imposition of the bill.

A year after the UVF's establishment, to fight off the UVF, the Irish Nationalists, too, created the IV. Moreover, the British government had attempted to pass a second bill in 1893 and a third Home Rule Bill in 1912 but unfortunately were not passed, as the majority voted against it.

Eventually, James rose to the rank of lieutenant in the Irish Volunteers. Then he was given the big responsibility of memorizing the serial numbers of Swiss bank accounts that the IV owned. He was tasked to memorize those serial numbers because of the risks that had they been written on paper, the British may have gotten hold of them, and he may have gotten captured. Thankfully, James had a sharp memory and memorized those serial numbers perfectly, and successfully sent them without being caught to Éamon de Valera.

The latter was prominent in 20th century Ireland as a statesman and political leader that became one of the top leaders of the rebellion and the President of the Free State of Ireland. The rebellion's motive was to free Ireland from Britain's control.

With the desire for freedom for his country, James was part of the rebellion. Later in 1914, with a plan of starting a revolution, he met the top leaders of rebellion with the likes of Padraig Pearse.

The IV, later, was able to purchase World War I German rifles in Germany with the details on the Swiss Bank accounts James relayed.

Unfortunately, they experienced some drawbacks in acquiring those rifles. Aside from the fact that Germany barely had any route for getting ships from Germany to Ireland. What made it more of a struggle was that there were British spies in Germany. Little did the IV know that the British spies caught them in action.

The British spies saw them carrying out crates of rifles and followed their trail, which led to Banna Strand, a beach located in County Kerry. Then on April 21, 1916, Roger Casement, a British nobleman, attempted to land a German U-Boat that carried the rifles for the Irish Republican Army (IRA), he was captured by the British. He was brought to Britain, where he was put on trial and executed with others that were captured that day.

Later, in 1916, James had returned to his home in Creeslough in County Donegal. It was not far from Doe Castle, where one of the main roads went to Letterkenny, a small city in County Donegal, Ireland.

Because many small roads connected to this major road near Doe Castle, the area became a hub that people radiated from. The convergence of these smaller roads at Doe Castle meant that travelers and locals would gather there before setting off to various destinations. This network of roads made Doe Castle a central point of movement, facilitating commerce and communication in the region.

When the Easter Rebellion started on Easter Monday, April 24, 1916, James became the commandant of the Doe battalion under the IV.

The Doe battalion only had 38 to 40 people. However, the Irish Republican Army were dirt poor and were usually farmers who needed more guns to fight against the British. So, to provide guns for themselves, they broke into the police stations to get records of all gun owners who had to be registered and broke into all those gun owners' homes to take their guns.

On their mission to acquire guns, a man in Creeslough, a doctor, whose name was Wilkinson—did not give up without a fight He fired his gun, and hit James on the shoulder. James was then transported to Dublin town via an ox cart. Back then, if they had a car, it might take them two days. By an ox cart, however, it took about five days to find a doctor who could be trusted to operate without telling the British.

James' brother, John, who was also living in Ireland and worked as a stone cutter was also quite an activist. He had had access to dynamite because of his work, and he planned to use it in attacking British troops by blowing up a railway car they were riding near Creeslough.

However, the dynamite barely blew up the railway car -- it had not been blasted out of the tracks -- however, unlucky for John, the police were on his tracks. How? John had lent a jacket to his friend that helped him blow up the train tracks. His friend left the jacket at the scene of the railroad track explosion. In his jacket's pocket was a letter to him with his address on it.

Because John was known for being an activist, the police came for him when the incident happened. They just took him and said, "Okay, you're coming with us."

The police also accused my great grandfather, James, of some crime. What it was exactly, we aren't certain. Maybe it was the same thing his brother, John, did. Both men were imprisoned in Letterkenny jail, along with ten other Irish prisoners. They were eventually transferred to Derry Gaol, also known as Londonderry Gaol, which was a notable place for the incarceration of the Irish Republican Army. It was out in a wild area, away from civilization.

They soon tried to escape. It was all planned. The prisoners' wives, who were allowed to visit, snuck in chloroform to help free their husbands. A car was going to pick them up as well. So the prisoners used the chloroform on the guards during the night shift for them to lose consciousness. Although it was not their intention, one or possibly both of them died, and the prisoners broke free. Or so they thought.

The car that was supposed to come to pick them up had not come. So, there they were, waiting for nothing. It was funny because they went back inside the jail and locked themselves up since they did not get picked up.

The new guards came for the next shift the following day, they saw the dead guards, and they immediately punished the prisoners. The guards established a system where they put black and white peas in a hat for the prisoners' punishment, and whoever picked a black pea would get executed. James was lucky enough to pick a white pea.

Unfortunately, a close friend of James had picked a black pea. His Friend, among many other inmates, was executed.

Britain soon suppressed the insurrection in Ireland. However, James, meanwhile, was in solitary confinement and essentially on death row. That is because the Civil War was still going on, and the British were still around and wanted him executed. James fought for the 32 counties of Ireland to be reunited and to be completely free from British rule.

However, the insurrection ended with the result of the freedom of Southern Ireland from the British, while only the six counties in Northern Ireland were kept under British control. James was also finally set free.

After he was set free, James went back to Creeslough. James built a two-story stone house in 1916. However, James' life was made miserable by the Black and Tans. They were the British constables at the time. They wore mixed uniforms in partially black and partially tan, which was the reason why they were called Black and Tans.

They put James under house arrest and would constantly mistreat him. They frequently visited his house and checked on him. When they would come, they did nasty things like kicking down his fences, running off his cattle and chickens, etc.

Around 1924, having grown tired of all the harassment he got from the Black and Tans, James went back to America. He did not have enough money to take his family with him at that time. It was after one year that he was able to send enough money for his family to come to America.

His family was then finally reunited in 1925 when his wife and children, Patrick Sr., James, and Kathleen, followed him to America, where they lived as a complete family in Chestnut Hill.

There even came some additions to their family, with three more children born in America: Cecilia, John, and Anna.

For a living, James started a partnership with his brother, Patrick.

They built nice, big stone homes, themselves, in the Wyndmoor area, which they sold. They did all the carpentry work. Or they at least did all the fitting, putting them into the walls, etc. They made about half a dozen homes until the two brothers had a falling out.

Moreover, in 1929, there, unfortunately, came the Great Depression. The devastating global depression in the economy caused by a drastic decline in prices in the stock market called the stock market crash. Since money was not flowing through the economy, people could not afford houses.

But things took a turn for the better during the time of World War II when the FBI came to James' Chestnut Hill home and whisked him away. Annie mistakenly thought it was because of his involvement in the Irish Rebellion. Actually, it was for a different reason. The FBI took him to the White House. It was President Franklin D. Roosevelt's term at that time. They wanted him to work on construction for them. He worked in the White House for a couple of years.

His work schedule would go something like this; on Monday, he would ride the train to go work in the White House, and he would come back home by Friday night. In the White House, he worked on excavating a mammoth room for Presidents and many officials to go to, in case the White House got bombed or in case of any other situation that would put the President and others in danger.

Later, he worked on other buildings in Washington

DC, where they had to handle gigantic pieces of stone. They had put these stones into place on most important government buildings where the facade is most important, and everything must fit. This continued into President Harry S. Truman's term.

During World War II, my grandfather, Patrick Sr., enlisted in the Navy, along with his brother, Jim. This occurred about two to three months after the United States entered the war following the December 7, 1941, attack on Pearl Harbor by the Japanese.

Around February or March of 1942, when my grandfather joined the Navy, he was offered the rank of Petty Officer Navy Machinist 3rd Class (PO3) due to his prior experience as a machinist. After graduating high school, he began his career as a tool room attendant, eventually mastering machine tools such as lathes, milling machines, and grinders to produce precision metal parts. His skills earned him a promotion to machinist.

Although Jim also had experience as a machinist, Patrick Sr. was more seasoned, and he too was offered the same position of Petty Officer Navy Machinist 3rd Class.

In the U.S. Armed Forces, enlisted pay grades range from E1 to E10. Patrick Sr. started at the E4 level and worked tirelessly to earn promotions. His dedication and knowledge propelled him from third class to second class, then to first class, and finally to chief petty officer.

Remarkably, he consistently scored the highest on every advancement test he took.

My grandfather was later assigned to the USS Auburn, a newly acquired ship. The ship had been moved to the Hawaiian Islands, so he was flown out there and became part of the ship's crew, called the Amphibious Group Command 10 or AGC-10. The ship was commissioned on July 20, 1944, under the command of Capt. Ralph Orsen Myers. At the time he was sent to be a part of the AGC-10, the USS Auburn, was being tested through sea trials. The crew had four admirals. They would land the ship often in Okinawa, Japan. The ship was one of many communication ships named after mountains in the United States. The USS Auburn got its name from a mountain located in the northwest of Cambridge, Massachusetts.

As a communication ship during World War II, the USS Auburn was equipped with huge antennas, advanced communications equipment, and extensive combat information centers. These facilities allowed it to receive and process information from ground forces, as well as coordinate with other ships during large-scale operations. By effectively relaying communications between land and sea, the ship played a crucial role in directing troops and resources to where they were most needed on the battlefield.

Additionally, the USS Auburn was armed with powerful guns capable of reaching targets 18 to 20 miles inland, providing significant artillery support to ground forces operating far from the coast.

The USS Auburn was an amphibious force flagship that was a Mount McKinley-class amphibious force command ship. It had been the flagship twice for two amphibious groups during large-scale operations in the Pacific. The first time was for the Commander, Amphibious Group 2, Pacific Fleet. On February 19, 1945, when the US underwent an assault on Iwo Jima, it took control over hundreds of Amphibious Group 2's ships.

The second time was for the Amphibious Group 5, when it joined in the last battle of World War II, the Battle of Okinawa. On May 31, 1945, it managed hundreds of the Amphibious Group 5's ships on Okinawa. At that time, the Japanese launched frequent air attacks, but it miraculously avoided damage. This last battle of World War II resulted in the defeat of the Axis Powers.

At the young age of 25 or so, Patrick Sr. was already Chief Petty Officer (CPO), Machinist's Mate. The CPO, Machinist's Mate oversaw all the chiefs below deck. The CPO boatswain is the senior enlisted rank of all above deck. My grandfather also became the person in charge of running the engine room and desalinating water for each guy on the ship to get their daily supply of a bucket of water for coffee, drinking, shaving, and whatever.

That was the highest rank he obtained in the Navy. When World War II ended in 1945, he planned to get out of the Navy, but he was offered the opportunity to be a warrant officer. He was not interested in the position because that position was neither a commissioned officer nor a non-commissioned officer. It was in between.

Instead, he chose to go back to his field of expertise before joining the Navy, working in a machine shop.

At a dance held by a non-profit charitable corporation, which provides all kinds of entertainment for the US Armed Forces called the United Services Organization (USO), Patrick Sr. met a lady named Alma. They fell in love, got married, and lived in Chestnut Hill, PA. Alma soon gave birth to a boy, who is my father, Patrick Jr., on March 11, 1944. Following Patrick Jr., they had seven other children, Alma, Michael, Joseph, Jim, John, Theresa, and Mary.

Money was tight until Patrick Sr. established a machine shop late in 1956 to early 1957 called the Universal Machine Shop. He rented an old belt-driven equipment machine shop located in the equivalent of a large two-story garage in Germantown. In 1958, he needed more space and equipment, so he went into a partnership with a man named Charlie Sheela in a more modern building with newer, better equipment. He kept the same name as his previous machine shop. Charlie had the money, and my great grandfather had the brains. Patrick Sr. had made approximately $50,000, which was quite a sum of money for a year or two in business. (a new car at that time cost $3,000).

Within two to three years, Patrick Sr. bought Charlie out. The business was then all his, and he renamed the shop McNulty Tool & Die shop. He converted his business from a machine shop to a tool and die shop. "progressive dies a specialty." The shop operated in the means of an assembly line, with each employee doing one specific task.

Patrick Sr.'s tool and die shop became exceptionally profitable, leading him to face high tax rates of 50%, 60%, or even 70%. He made more profit and had better ways of doing things than most other machinist shops. The government found it unusual that someone with only a 12th-grade education was earning such substantial income and scrutinized his financial reports. As a result, they imposed significant taxes that greatly reduced his earnings.

In response, Patrick hired an accountant, and from that point on, he no longer had problems with the government.

An example of my grandfather's ingenuity would be the collapsible safety ladder that would roll out on its own, with an adjustable length.

This product of his was a solution he came up with for emergencies like fire in apartment buildings. It was beneficial and in demand for those who were living in apartment buildings.

Escaping a fire can be a struggle; not to mention terrifying.

Firemen can come to the rescue with their hook and ladder fire trucks, but their ladders can only reach up to the fourth floor. Also, they may have pike poles with them, which was like a stick with a hook on its end that they would use for pulling something out, like a window, to enter a burning place, but it can only be as long as 10 feet.

So for the safety of the residents of apartment buildings, the local government of the city of Philadelphia passed this ordinance that required that apartments for rent should only be located on the first to third floors because hook and ladders cannot reach higher than that and that they also had to have an exit.

As the ordinance had been passed, Patrick Sr. developed this collapsible ladder that was designed in a way that it would be mounted on the apartment wall inside going to the outside. It just had to be thrown out to use it, and it will roll out by itself. In the design of his ladder, it had an upside-down "U" at the end, wherein another set of chains can be added to make it longer if needed.

His safety ladder had an excellent, convenient design. It was one of only two safety ladders that gained approval from the City of Philadelphia for official use in apartment buildings. He was able to earn a fortune when he came up with his safety ladder design and manufacturing those safety ladders.

I believe my great grandfather, grandfather, and father share three important values: being family-oriented, always striving to be number one in what they do, and having a sense of duty toward your country.

Apart from that, another thing those three generations of McNultys have in common is signing up for military service. My great grandfather fought for the Irish to be freed from British rule. My grandfather fought as a strike back at Japan for attacking the United States. In my opinion, my father fought in Vietnam because of his belief that everyone should donate at least two years of their life to serve in a country as great as we have.

CHAPTER THREE

The Domino Theory & the U.S. Involvement in the Vietnam War

This chapter reflects what I understood from my father's perspective about how the Vietnam War started. It may not align with how others have perceived the events, but it represents his account and interpretation.

During the Cold War—a long period of tension between capitalist countries led by the United States and communist countries led by the Soviet Union—Vietnam became a key battleground. The conflict was between pro-communist North Vietnam and anti-communist South Vietnam. The U.S., along with allies like Thailand, Australia, New Zealand, and the Philippines, supported South Vietnam.

In contrast, North Vietnam received backing from communist nations such as the Soviet Union, China, and North Korea. Motivated by the Domino Theory, the U.S. government felt compelled to intervene by sending troops to Vietnam.

The Domino Theory was the U.S. government's reason for preventing the spread of communism. It suggested that if one country in a region fell to communism, neighboring countries would follow, much like a row of dominos falling one after another. The U.S. feared that the expansion of communism would threaten its alliances and undermine the American way of life.

President Harry S. Truman first introduced this theory in the 1940s, particularly concerning Greece and Turkey. It gained widespread attention in the 1950s when President Dwight D. Eisenhower applied it to Southeast Asia, especially Vietnam.

From my father's perspective, the Domino Theory was "just crazy stuff." He argued that if South Vietnam fell, North Vietnam would dominate, leading to a chain reaction where countries like Cambodia and Laos would also succumb to communism. To him, losing one region to communism meant the potential loss of all surrounding areas.

Before U.S. troops supported the South Vietnamese army, Vietnam was already facing internal conflicts that eventually escalated into the Vietnam War.

In the early 19th century, Vietnam was colonized by France and later invaded by Japanese forces during World War II. A pro-communist leader named Ho Chi Minh sought independence from both French and Japanese control. In 1941, he established the Viet Minh (League for the Independence of Vietnam) to fight against these occupying forces.

When World War II ended in 1945 with the defeat of the Axis Powers, including Japan, Japanese forces withdrew from Vietnam, ending Japanese control. However, the French remained and installed Emperor Bao Dai—a French-educated leader—to govern. Unlike Ho Chi Minh, who promoted communism, Bao Dai preferred to maintain close economic and cultural ties with the West.

Ho Chi Minh and his Viet Minh forces took control of North Vietnam, declaring Hanoi as the capital of the Democratic Republic of Vietnam. In response, France, under Bao Dai's authority, established the State of Vietnam in the South with Saigon as its capital, marking the beginning of the First Indochina War in 1946.

In 1954, North Vietnam achieved a decisive victory over the French at the Battle of Dien Bien Phu. That same year, the Geneva Conference was held in Switzerland to address conflicts in Indochina. Nine parties attended, including North and South Vietnam, the Soviet Union, China, the United States, France, the United Kingdom, Laos, and Cambodia.

The conference resulted in the Geneva Accords, which included military agreements, declarations from each party, a plan for reunification elections, and the Final Declaration of the Geneva Conference.

One significant outcome was France's withdrawal from Vietnam following its defeat. Vietnam was temporarily divided at the 17th parallel: communist North Vietnam under Ho Chi Minh and anti-communist South Vietnam under Ngo Dinh Diem, who succeeded Bao Dai. This division was intended to last two years until national elections could be held to reunify the country under a single government.

However, U.S. officials opposed the Geneva Accords. Only France and North Vietnam signed the agreement. The U.S. feared that if national elections proceeded, South Vietnam would lose to Ho Chi Minh, leading to a fully communist Vietnam. To prevent this outcome, the U.S. blocked the elections and reunification.

Using the Domino Theory as justification, the U.S. supported South Vietnam by establishing a new anti-communist government and providing financial and military aid, including training and support for the South Vietnamese Army. This intervention set the stage for the Vietnam War.

According to my father, the leaders in South Vietnam misused the aid money provided by the U.S. Corruption was widespread, and their government was overthrown multiple times. This corruption further convinced the U.S. that South Vietnam was destined to fail. The U.S. installed another leader, but he proved to be just as corrupt as his predecessors.

With no elections or reunification taking place, North Vietnam began guerrilla warfare against South Vietnam in 1957. The conflict in South Vietnam was fought between the anti-communist South Vietnamese army, supported by U.S. troops, and the Viet Cong—a communist group allied with North Vietnam and based in the South.

When Vietnam was divided, similar to the division of

North and South Korea after World War II, the U.S. tried to manage the situation. While they had more success in North Korea, in Vietnam, they decided to take direct control of the fight themselves.

However, there was no solid reason for U.S. involvement in Vietnam. Although the Gulf of Tonkin Incident—also known as the USS Maddox Incident—became the trigger for U.S. escalation, reports and publications like the Pentagon Papers and statements from officials like Robert S. McNamara suggested that the incident was fabricated to justify expanding U.S. involvement.

Nevertheless, the incident led to the U.S. Congress passing the Gulf of Tonkin Resolution on August 7, 1964, which authorized the U.S. to "take all necessary measures to repel any armed attack against the forces of the United States and to prevent further aggression" from North Vietnam.

Common accounts of the Gulf of Tonkin Incident state that on August 2, 1964, the destroyer USS Maddox was conducting a signals intelligence patrol in the Gulf of Tonkin (now known as the East Vietnam Sea) when it was allegedly attacked by three North Vietnamese torpedo boats. The USS Maddox fired back and damaged the torpedo boats. A second alleged attack occurred on August 4, involving both the USS Maddox and the USS Turner Joy. They claimed to have spotted torpedo boats on radar, although no visual confirmation was made. While the August 2 attack was documented, the August 4 attack was later questioned and deemed unlikely to have occurred.

My father believes the incident was fabricated, arguing that the U.S. went to war on false pretenses. He noted that while some believed there was only one attack, others contended there were two, and some even suggested there was no attack at all. He felt that the U.S. might have been reacting to vague threats and then decided to escalate the situation into a war without just cause.

According to him, if someone had genuinely attacked our Navy ships, there would have been clear evidence like bullet holes or damage from cannon fire. However, no such proof was provided.

This is how I understood the events from my father's perspective. It may not align with other accounts of how the Vietnam War started, but it represents his experience and understanding of that time.

CHAPTER FOUR

The Pursuit of a Career in the US Armed Forces

To become a soldier fit for battle, a man or woman who aspires to be in the US Armed Forces must attend a series of schools and rigorous training of all sorts to prepare him (or her) for going into war. My father's first step in his pursuit of a career in the US Armed Forces was joining the Army Reserve Officer Training Corps (ROTC).

ROTC is a program offered in colleges and universities for young adults interested in joining the military to be prepared to assume the role of a commissioned officer. My father, Patrick John McNulty Jr., pursued the Bachelor of Science in Industrial Management course at La Salle University while attending the ROTC that was also offered there.

Because La Salle University was getting money from the government, ROTC was mandatory for the first two years of college, then optional for the last two years. So, every student there attended ROTC for their first two years.

Then my father joined ROTC Summer Camp in 1965. His courses were held both in the field and in the classroom. My father said ROTC was nothing big, though, and described it as just marching and stuff.

Since my father's third year of ROTC in 1964 was an elective, he had decided to try to get into a prestigious military school such as the United States Military Academy at West Point, Naval Academy, or Coast Guard Academy.

Getting appointed to any of those schools is not that easy, though. The fate of the men who dreamed of getting admitted to those schools lies in the hands of each senator. Each of the senators got to choose among the many aspiring applicants—just one person each year—to go to those schools. To get appointed, my father had to send letters of recommendation to the senator.

So, my father got letters of recommendation from nuns and teachers to be sent to the senator. Upon acquiring the letters, he sent them to the senator with much hope. Apart from that one person whom a senator would choose, there were also backups. There was a primary candidate, then a first, second, and third alternate. Alternates were also considered to serve as replacements if the first choice fell short in some of the requirements for being selected.

Supposing the first choice's score in the college exam was not high enough, or if his physical condition did not comply with the physical standards, he would lose his chance. The opportunity would be passed on to an alternate choice and so on until someone more fit was found.

My father was set to quit college even up in his third

year to go into the military at West Point or something, if it worked out. Unfortunately, my father was not selected but was only the first or second alternate.

It did not reach him, so the other guy went to West Point. Because he was not selected, he continued his studies at La Salle University and graduated in 1966. When he graduated and finished ROTC, he finally got commissioned as a 2nd lieutenant. After that, he had been working at my grandfather's tool and die shop for a while.

Alongside the developmental process of my father as a military man and his work as a machinist for his father, love also blossomed between him and a good woman that is none other than my mother, Alice Ann, whom he married on August 27, 1966.

Following ROTC, my father attended Field Artillery School from September to November 1966. Field Artillery School had prepared him for some action with fire support systems out in the field. As a forward observer, one of his primary roles was to direct the troops manning the artillery at the firebase (FB), where they would fire the artillery to support the infantry.

It was a huge responsibility he had in his hands because while he had to make sure he was directing fire toward the enemy with precision, he also had to make sure that his fellow troops were unharmed in the process. Artillery is so powerful that it can put out fire in greater amounts and much faster compared to the infantry. It can also kill many more of the enemy and even wipe them out. They say in the military that the infantry is the queen of battle, and artillery is the king of battle. As Fredrick the Great of Prusia said, "The artillery lends dignity to battle. Otherwise, it would just be a vulgar brawl." Hand in hand, the artillery and the infantry fight against the enemy with their distinct roles and responsibilities. As my father recounted, in his first fire mission when he officially became an FO in Vietnam, he said he felt scared, and he was cautious not to screw up. Thankfully, the Field Artillery School equipped him with the know-how in working with artillery.

Field Artillery School is attended by soldiers, officers, and Marines to train fire support systems. In the Field Artillery School, the field artillery mission is instilled in the students. The field artillery mission is to destroy, neutralize, or suppress the enemy by cannon, rocket, and missile fire and help integrate all lethal and nonlethal fire support assets into combined arms operations. They were trained on how to deal with fire support systems when they go out there to face the real thing at war. They were taught the tactics, techniques, and procedures to be done when using fire support systems. They were also given individual training on the basic level and mid-level leadership skills.

It was in the winter of 1967 when my father attended Ranger School. He was turning 23 years old at that time. It all began in Fort Sill, Oklahoma, where they were asked questions like, "What would you like to do?", "Do you want to go to Airborne school?", "Do you want to go to

Ranger school?", "You want to go to Pathfinder school?", "Or would you rather do anything else?" and "What would your duty stations be? List four of them."

My father knew exactly what he wanted to do, and that was to go to Ranger School. He had always been awestruck by the Rangers' reputation of being the top elite as like what he had seen in the movies about the Rangers from World War II.

He so looked up to the Rangers, that he thought, "Why not go through the training that they had and be one of the best there is?"

Ranger School was so realistic; it had an old wives tale that it was the only Army school that was authorized a death rate, but this was not correct.

During the Vietnam War era, Ranger School did see an increase in training intensity to prepare soldiers for the harsh conditions they would face in combat. This intensified training could have contributed to higher injury rates, but there was never an official policy that accepted or authorized a certain number of deaths.

It's possible that this myth arose from misunderstandings about the inherent risks involved in such strenuous training or from isolated incidents that were misconstrued over time. The military continually reviews and updates its training protocols to enhance safety while still achieving training objectives.

So, he put going to Ranger School on his dream sheet. Then off to Ranger School he went, knowing that he had a 90 percent probability of being sent to Vietnam, which was two years into the official Vietnam war.

The prestige of a ranger that my father admired came with a price. The kind of training offered in Ranger School is one of the toughest among the different military training courses available. The course involves rigorous training that pushes both their minds and bodies to the limits. It develops students, whether officers or enlisted soldiers, into combat-ready soldiers equipped with functional skills in combat arms and excellent leadership skills.

Ranger School's training was hard enough, to begin with, adding to its challenges. The time my father joined Ranger School was the cold winter temperatures. Given that it was winter at that time, many people did not want to go to Ranger School because all the activities done in Ranger School, whatever the season, are done the same, only the difference is it is freezing cold out there during winter. Not a lot of men were up for doing the strenuous tasks that are expected of them to complete out in the cold. On the other hand, my father was determined to undergo the course and graduate.

It would be so cold when they were out training that they even called Ranger School in the winter season "frostbite six." Picture being out in the mountains amidst freezing temperatures doing rappelling.

Another example is when they did the slide for life, wherein they would walk on top of telephone lines around Victory Lake from one telephone pole to the next. If one fell off, he would fall into the ice-cold, frozen lake with about one to two inches of ice. The ice would break when he fell and landed on it, putting him into freezing water.

The navy had underwater demolition teams (UDTs) in wet suits in case someone got stuck in the lake bottom after dropping into the lake from a height of 40 feet above.

From what my father experienced in Ranger School back in 1967, Ranger School at present is quite different. Its system has changed over time. We found out that to date, the course runs for eight weeks or 56 days. When my father went to Ranger School, it was nine weeks or 63 days. There, they went through three phases: the Benning Phase, Mountain Phase and Florida Phase.

Apart from the exhausting activities the school had them do as part of their training, it also upheld a strict diet and sleep hours. The continual lack of food and sleep just made everything more challenging. The reason for the strict regulation of diet and sleep is because out in combat, food, and rest are the least of the infantry men's priorities. Security, weapons maintenance, and personal hygiene are more prioritized. Hence, even just at Ranger School, soldiers are already being trained on their food intake and sleep.

When my father attended Ranger School, the trainees were given only one meal per day—a ready-to-eat ration—and were allotted just two hours of sleep each night over a period of 63 days. Surviving the full nine weeks of such intense training with so little food and rest meant that participants would typically lose around 40 pounds. The training was extremely rigorous, and the limited food supply was insufficient for students burning a massive number of calories daily. On average, they lost about two-thirds of a pound each day.

The Army became concerned about the significant weight loss among Ranger School students and consulted a dietician to investigate the issue.

The dietician calculated that someone undergoing Ranger School training required approximately 4,800 calories per day due to the physical demands. However, a C-ration provided only about 1,200 calories. Given that they were burning 4,800 calories daily and consuming just 1,200 calories, and considering that a pound of body weight equals 3,500 calories, it was inevitable that they would lose weight.

My father recalled an instance when someone asked him, "Why are the men in Ranger School losing so much weight?" He replied, "How can you expect them not to lose weight? They're burning up to 4,800 calories each day while their meals provide maybe 1,200 calories at best. It just doesn't add up."

The guy who asked my father that question regarding the men's weight loss in Ranger School joined the Army 20 years after my father did. By the time that guy went to Ranger School, the school had increased the diet to one and a half meals as well as the hours of sleep to three hours. Then 20 more years after that, there was again an increase in the diet, making it two meals a day and in sleep, making it four hours of sleep. My father noted that if calculated, there is a considerably big difference in the days of actual training between the students who got two hours of sleep versus the students who got four hours of sleep at night in nine weeks or 63 days.

So, for those who attended Ranger School in 1967, they only had two hours of sleep: 63 days in training x 2 hours of sleep = 126 hours of sleep. When subtracted to 1512 hours in 63 days, then divided by 24 hours in one day, the result is they get 58 days of training. Then if the 126 hours of sleep are doubled, making it 252 hours, and then applying to it the same calculation as earlier, that is fifty-three days of training, which is an extra five days for the former.

That means compared to the students who got four hours of sleep, those who got only two hours of sleep essentially got an extra five extra days of training on top of the 63 days.

Ranger School is the closest thing to actual combat. It is that realistic. It goes on for nine solid weeks, wherein students train for 22 hours a day with only one meal a day and only two hours of sleep.

Everything else was like in the real thing. Like I said earlier, Ranger School has changed over time. However, I can share my father's own experiences from Ranger School as he recounted.

As a means of safety and precaution, like in case, students needed to be rescued, parachute jumpers (PJs) and even Marines and navy seals were present. In my father's opinion, though, the navy seals were probably in Ranger School so that they can get a nine-week vacation.

The first phase called the Benning Phase, also known as the "Crawl Phase", took place in Fort Benning, Georgia, in two different camps, Camp Rogers and Camp Darby. The phase assesses and develops the students' physical and mental skills required to complete each phase in Ranger School and to be established in them as a soldier. As my father recalled, in the first three weeks or so, they did plenty of physical tests, hand-to-hand combating, duck-walking around the football field, and there was also the swim test.

The swim test is a tough challenge for the weak swimmers and even more so for those who cannot swim. The Ranger School tests are challenging, but the swim test was super hard for those with little to no swimming skills. In the swim test, they had to enter the water twice. The first time was getting themselves wet and standing around in the freezing temperature for 30 minutes. Next, they climbed up to a high dive, where they walked off and jumped from while they were blindfolded. Then they swam a distance of 30 feet with their rifle.

If they dropped their rifle to survive as they tried to finish the 30 feet-long swim, they were classified as a weak swimmer, but at least they made it. They would also be paired up with a ranger buddy that was not a weak swimmer. Those who cannot swim had it worse because since they cannot swim, they just dived down the water, held their breath, walked the whole 30 feet underwater, and then got up when they reached the end. That was how great their motivation level was not to get dropped from the course.

They also had this morning routine in the first three weeks to get up and run five miles. While they did the five-mile run, the Ranger instructors set up this icy path. The instructors would create this sea of mud using fire hoses with hot water, which would turn into an icy path that just froze up with winter's cold temperature by the time they got back. The instructors also put barbed wire on it for them to crawl under. When they came back from their five-mile run, the icy, barb-wired path was all set for them to go through. They plunged through this obstacle course, crawling on their bellies with their rifles in hand.

While they were at it, they tried to keep their hands dry. Otherwise, if their hands got muddy, their rifle might slip from their hands and crack the surface of the ice, and they would fall six feet deep into the water. They had to keep their hands dry because after that; they did overhead bar runs on monkey bars that had unpinned bars, so the bars could rotate in their hands.

In the monkey bars, 10 people at a time would be made to do Ten chin-ups. When they were up there, the ranger instructors would say things like, "Ranger number two, you better get your head up there!", "Ranger number four, what is your problem?!" So, it was not just, "Do 10, and get off". The instructors were hateful towards them, and they were made to wait as long as possible.

However, in the summertime, the instructors have a different kind of challenge in store for them in the water. The instructors put them through the challenge of wading through the water where snakes have been put in. Then, whether summer or winter, the next thing they had to do was climb a runged telephone pole and transfer from one telephone pole to another, which had a distance of about four feet between them, by walking on the telephone lines. During winter, the water would be icy cold, and they had to be careful not to fall into it. If they did, bad luck.

There was also the competence test in Victory Lake, where they had to climb up this eighty to a hundred feet high steel "I" beam with re-rod welded across that served as ladder rungs going up to this five-foot square platform. Just going up can be frightening, especially for those afraid of heights. Adding to the fear factor was the absence of any safety equipment. They had to be careful climbing up then.

When they reached the platform, they grabbed hold of this steel pulley, put their chin up, jumped off, and went on the slide for their life down an almost 300-foot ride terminating in the water, where navy UDTs broke the ice formed in the lake from the cold winter temperature so they could drop in. They had to grab on tight to the rope because if their grip slipped, it was almost a hundred feet drop down to the water.

When they were about to approach the water, they had to take precautions before landing. They had to lift the lower half of their body and position it parallel to the water before letting go.

Otherwise, their legs would be dragged down and hit the water, and the incredible amount of force would rip them right off the rope as they plunged in. The navy UDTs were also on standby if they needed rescuing.

Navy swimmers were there, too, in case someone got knocked out or had to be retrieved out of the water.

There were a bunch of other competence tests they underwent in the Benning Phase. Then they moved them up to the north Georgia mountains in Camp Frank D. Merrill in Dahlonega, Georgia, to move on to the Mountain Phase, also known as the "Walk Phase."

This phase got them familiarized with working in the mountains, which included military mountaineering tasks and taking control over squads and platoons in continuous combat patrol operations amidst the mountains.

On 90-foot cliffs, they learned how to do rappelling, which is descending from cliffs by sliding down with a rope secured at the cliffs' top surface and attached to the waist harness. They also learned how to climb up the mountains during nighttime. They would drive in pitons in cracks or seams on the surface they rappelled or climbed with a climbing hammer. Pitons are metal spikes that served as anchors both for assistance in climbing and protection from falling. They would then put snap links on it to secure the rope they would extend. Those kinds of activities went on for three weeks.

They also underwent patrolling for ten days. Their company captain that led their class in the patrolling activity was an Army captain.

Their class was composed of 300 students under the military's different branches, such as the navy seals, air force para jumpers, Marines, and Army. Everyone in the class had their ranks taken off their uniforms, but they knew that the man leading them was an Army captain and that he had been a Rhodes scholar. However, with the very low temperature since they were out patrolling in the snow, the physical and mental strain that could be experienced during training, etc., their company captain had a mental breakdown. He was not acting nor thinking like an officer, but rather, he declined to the level of a three-year-old kid.

The winter temperature could be so harsh on them that people burned their maps to make themselves warm. Unfortunately, those who no longer had maps because they burned theirs would lose points when the ranger instructors asked them where their map is.

Then, finally, they moved down to Camp Rudder,

Eglin Air Force Base in Florida for the Florida Phase, also known as the "Swamp Phase" or the "Run Phase." They were taught about waterborne operations, small boat movements, and stream crossings in this phase.

This phase also helped them enhance their abilities in planning and leading in airborne, air assault, small boat, ship-to-shore, and dismounted combat patrol operations. Also, they underwent extended platoon-level operations exercises that develop their combat arms functional skills amid mentally and physically stressful conditions that took place in a swamp environment.

My father recalled always being in the water for about a week. Never were they on dry land. In one of the trainings they had, called the Inflatable Boat Training, they rode boats four to five miles down the Indian River, then they jumped out of the boats, and off they went to the swamps. Each of them had a tiny navy one-cell flashlight for safety purposes because there were sinkholes in the swamps that may have been about 20 feet deep. The flashlights were for alerting the other men if they slipped into a sinkhole. They just had to turn it on so they could be found and rescued.

They would spend the entire 24 hours of a day in the swamps. When someone encountered an obstacle, everyone stopped for a few seconds.

If anyone slept longer than the expected hours of sleep, he would be surprised to wake up all alone. He would also not have luminous tapes attached in the front and the back of his headgear, called the patrol cap, and would have to quickly take off to catch up with the the noise of the men ahead of him.

Among all the schools the Army had, Ranger School was the toughest. The navy seals would even use it as the base for their training.

They would go to Airborne school first, Ranger training next, and build from that.

After all the hardships my father went through in Ranger School, he passed with flying colors and graduated in March 1967, belonging in the top one-third of the class of 300 students who made it.

Unfortunately, one-third of the class experienced injuries such as broken backs, arms, and legs and quit because of that. As my father got through, he was given orders, and five months later, he was off to Vietnam.

In August 1967, he officially volunteered for service to go on active duty and five years in the reserve. Because my father left for Vietnam, he was only able to spend life as a married man for about a year, and he only got to experience being a father to me, his son, for four days before he had to leave for Vietnam.

CHAPTER FIVE

The Unsung Heroes: REMFs in the Vietnam War

Back in the thick of the Vietnam War, labels had a way of sticking. One that was thrown around often, sometimes with a hint of disdain, was "REMF"—Rear Echelon Mother Fucker. Front-line soldiers like us used it to refer to those who weren't knee-deep in the mud and blood alongside us. It was easy to think of them as sitting comfortably behind the lines, untouched by the horrors we faced daily.

In those days, we didn't consider what those behind the scenes were enduring. It was simpler to assume they had it easy, but as time passed, I reflected on those tumultuous times, I began to understand how misguided and narrow-minded that perception was. This chapter is about setting the record straight and honoring the indispensable role that REMFs played in our shared history.

I remember the suffocating heat of the jungle, the air thick with humidity and tension. Our platoon was deep in enemy territory, low on supplies, and morale was wavering. We'd been on patrol for days, the weight of our packs matched only by the weight of uncertainty hanging over us.

When things seemed most threatening, a resupply helicopter appeared on the horizon. Despite the risk of enemy fire, the pilot set down in a clearing barely large enough to accommodate the chopper. Out jumped a crew of REMFs, rushing to unload crates of ammunition, rations, and medical supplies. They stayed just long enough to get the job done before lifting off, disappearing as quickly as they had arrived. They were our lifeline, a realization that would slowly challenge the misconceptions about them.

Those pilots and crew members faced dangers every bit as real as ours. Flying over hostile territory, they were prime targets for enemy gunners. Many never made it back, shot down while performing missions critical to our survival. Their courage was undeniable, yet we seldom acknowledged it at the time.

Then, the medics and nurses were stationed at field hospitals like Chu Lai. I can still see the sterile white tents contrasted against the backdrop of chaos. Wounded soldiers arrived hourly, sometimes by the dozens, each one fighting a personal battle between life and death. The medical staff worked tirelessly; hands steady even as artillery boomed in the distance. They stitched up wounds, administered blood transfusions, and offered words of comfort that often meant as much as medicine. These REMFs carried the emotional scars of war, haunted by the faces of those they couldn't save. Their battlefield was one of suffering and solace, where the cost of war was counted in human lives.

Intelligence officers, too, played a pivotal role that often went unnoticed. They spent countless hours analyzing aerial photographs, blocking enemy communications, and piecing together scraps of information to predict enemy movements. Their diligence allowed units like mine to avoid deadly ambushes or strike strategically when the odds were in our favor.

I recall a mission where we were rerouted at the last minute based on fresh intel. Later, we learned that the original path had been laced with booby traps and covered by enemy snipers. The REMFs had saved our lives without ever firing a shot.

Communications specialists were another group whose contributions were invaluable. In the dense jungles of Vietnam, where visibility was limited and danger crept behind every tree, reliable communication was our safety net. Radio operators maintained critical links between units, coordinated air support and facilitated medevac evacuations. I remember nights when the airwaves crackled with urgency, every message a piece of a larger puzzle. The men handling those radios worked under immense pressure, fully aware that miscommunication could have fatal consequences.

Even the clerks, cooks, and mechanics—often the subjects of our misguided jokes—were essential cogs in the machine. After days of cold rations, a hot meal could revive spirits in ways that defied logic.

Mechanics toiled in makeshift workshops, keeping our vehicles and equipment operational despite the harsh conditions. They improvised with limited resources, embodying a creativity born of necessity.

Administrative personnel managed the endless paperwork that kept supplies flowing and ensured that letters from home reached us. Those letters were a lifeline to another world, reminding us why we were fighting and what awaited us beyond the battlefield.

It's important to acknowledge that the war zone blurred the lines between front-line soldiers and rear-echelon personnel. Nowhere was genuinely safe. Bases were subject to cannon attacks, and enemy infiltrators could strike without warning. REMFs faced the constant stress of potential danger while performing their duties, often without the camaraderie that sustained those of us in combat units.

Moreover, the psychological toll on REMFs was profound. Many grappled with guilt, believing their contributions were lesser because they weren't on the front lines. They carried the burden of being labeled as less courageous or essential, an unfair and unfounded stigma. The reality is that war demands a collective effort, a mosaic of roles that together form the bigger picture of military operation. No matter how seemingly insignificant, every job played a part in the larger mission.

The aftermath of the war brought its own set of challenges. Many REMFs returned home to a society that didn't understand the complexities of their service. They battled the same demons—nightmares, flashbacks, a sense of dislocation—that haunted combat veterans. Yet, their struggles were often overlooked in the national discourse about the war. Despite their invaluable contributions, their stories were drowned out, leaving them in the shadows of history.

As I pen this chapter, I am struck by the realization that heroism wears many faces. It's found not only in the soldier who charges a hill under fire, but also in the one who ensures that the soldier has bullets in his gun and food in his pack. It's in the pilot who flies through hostile skies to deliver supplies, the medic who binds wounds with steady hands, and the mechanic who keeps the wheels turning against all odds.

The term "REMF" should be reclaimed, stripped of its derogatory undertones, and recognized for what it truly represents: the backbone of military operations. These men and women exemplified dedication, courage, and resilience. They embraced roles that were crucial, yet often thankless, driven by a sense of duty that transcended personal glory.

In reflecting on my own experiences, I recognize that my survival and that of my fellow soldiers were inextricably linked to the efforts of the REMFs. Their contributions were the invisible threads that held together the fabric of our operations. Without them, the war—and our place in it—would have looked very different.

I hope that future generations will understand the full scope of what it takes to fight a war like Vietnam. Celebrating the front-line heroes is not enough; we must also honor those who worked tirelessly behind the scenes. Their stories deserve to be told with the same reverence and respect.

In the end, it's clear that REMFs weren't just participants in the war—they were the unsung heroes who made victory possible. Without them, the wheels of the war machine would have ground to a halt. As time marches on, let their legacy be one of courage, sacrifice, and unwavering commitment to the mission, for their heroism is a vital piece of the story we cannot afford to forget.

Acknowledging their sacrifices enriches the narrative of our shared history and deepens our understanding of what it truly means to serve. One truth stands out: while some carried the fight, others took the weight of making it possible. And sometimes, that quiet, unseen courage is the hardest to recognize, but the worthiest of honor.

CHAPTER SIX

The Life of a Forward Observer: Courage, Precision & Survival in Vietnam

As Forward Observers (FOs) are on the front lines in war, one of the requirements in the FO selection process, according to my father is, "Being a self-thinking man, who through his thought process, is capable of dropping to the ground when an explosion goes off nearby or when bullets are fired in their direction." That would be the essential requirement for a soldier to have to qualify as an FO.

Apart from that, also required, as my father said, are of course, "A pulse and being of junior rank of the selector." My father, being a capable and an eligible commissioned officer that he was, became an FO for Delta Company, 3rd Battalion, 8th Infantry in 1967.

Very determined, my father aimed to be the best FO out there. After all, this was his dream. He prepared for this—from Reserve Officers' Training Corps (ROTC) to Field Artillery School to Ranger School.

Finally, he was out in the real thing. He went through rigorous preparations for it, and his time to serve for the

US Armed Forces finally came. He was assigned from 1967-1968 in Dak To, an area near the tri-border of Laos, Cambodia, and South Vietnam.

There had always been a shortage of FOs, with soldiers getting wounded or dying in battle around that time. Soldiers carried out their military service until they were injured or killed in action or completed their one year tour of duty. My father came in as a replacement to Delta Co, 3rd Bn, 8th INF's first FO, Lt. Larry "Shot-in-the-Back-of-the-Neck" Skoglund. Delta Company had been formed a month before my father replaced Larry.

Larry got wounded while he was doing his job out in the field as an FO. The unfortunate event was just an accident. He was shot in the back of the neck by a fellow infantry man. How did it happen? Well, it was because that infantryman did not follow Delta Co, Ranger Qualified Officer, Capt. Terence M. Bell's policy on chambering a round in their rifles. As my father detailed it, Bell's policy was that they never chamber a round in their rifle unless they find themselves in three particular situations. The first situation was if they were walking point. The second situation was if they could feel eyes on them. For example, if they emerged from a forest, into a stream with a clearing across it—an area in the forest that had been rid of trees, usually by B52 bombers, since trees in the forests of Vietnam could be as high as 120 feet—and they had to cross that open area that is maybe around 100 yards before re-entering a forest. In a situation like that, when they crossed the kill area one person or two persons at a time, while on the run and they could feel eyes on you, they could chamber a round until they re-entered the forest.

The third situation would be if they were being shot at. My father remarked that it seemed like a long-winded explanation about nothing. Ah, contrair! Adding that not all of the units he served with followed the same policy. Because someone did not follow that policy that one time when an infantryman was walking point, the inevitable happened.

What usually happens when walking point is that infantrymen would take turns. After an hour or so that an infantryman had been up in the front, he would get replaced and move back to the rear. So, the manner of the task is in rotation. However, one infantryman who rotated to the back did not clear or remove the live round from his rifle after rotating off point. Then, accidentally, he tripped on a vine, which caused his weapon to discharge, hitting a tree, and then the bullet ricocheted into Skoglund's neck. They called a medevac chopper to take Skoglund to the hospital.

Moments before the medevac arrived, Skoglund's radio telephone operator (RTO), Dale Young asked Skoglund, "Sir, I don't know if I can ask you or not, but since you're going in, can I have your canned peaches?"

Rooting from that accident when Skoglund got shot, had developed a friendship between my father and Skoglund. Thirty-five years later, at one of the earliest 3rd

Battalion 8th Infantry reunions, my father heard a familiar phrase, "Shot-in-the-Back-of-the-Neck." For the first time, he had met Skoglund. They found that they had much in common and also gained a newfound friendship, which then lasted for 15 more years.

Some troops can be stubborn, and not all units my father served with followed the captain's orders. Bell also had this rule that they should always make or cut their paths. Again, not all followed this rule. The idea adapted from Ranger School that "Only dead men walk trails," would be one of the reasons why some would not follow that rule.

Because they learned in Ranger School, "If you take the easy way, you'll pay!" My father explained it this way, "That's how Rangers earned their reputation. Whenever possible, they would use the most unfavorable conditions to their advantage: operate at night, when others want to sleep; operate during storms, rain, or bad weather; do not walk down trails, it's just a speed march to an ambush or worse, death.

They would take the least likely route to an objective because it's difficult. Most people opt for creature comfort, so it's less likely you'll meet them and risk compromising your mission." Another reason behind the non-compliance of troops may probably be because going out searching to make contact with Charlie can be exhausting.

For one, walking point was a physically and mentally tiring job. Infantrymen walked 20 yards ahead of the company, serving as the early warning group. They would just walk for miles, searching for Charlie. It takes a lot of courage to put themselves out there, which gave them the technical term, "Bait."

On top of that, it was a virgin forest with "wait-a-minute" vines scattered all over, obstructing their path and allowing them only a range of vision that is typically less than five meters. So, it was a struggle, and having to whack their machetes away at vines was an added burden.

My father's service as an FO for the US Armed Forces in Vietnam officially started in September 1967. When he and his fellow soldiers arrived at the Republic of Vietnam (RVN), the first thing they did before they went to the war zone was they went to Cam Rahn Bay, where they had their name tapes and patches sewn on their uniforms.

After that, they would be dressed for battle, hitting the field wearing a nice uniform with their Ranger tab, rank, and crossed cannon patches. It is quite ironic because when they would go out to the field, they would just be removing the patches from their uniform, as those patches would give Charlie a clue on who to target.

Through the patches, Charlie would be able to identify who are those of high rank or importance, and they are the ones Charlie would try to kill first.

After my father was all set, he was flown out and escorted to a firebase. A helicopter dropped him off in a landing zone (LZ) 200 yards away from the firebase.

Then someone escorted him from the LZ to the firebase (FB). Making his way to the FB, there was this fallen tree that he was told to take a large step to get up to and take another large step to get down from, which had to be crawled over. That was because buried under that tree was the dead body of a North Vietnam Army (NVA) soldier.

The body was not buried too deep, and the ground he stepped on was squishy. Nevertheless, as he finally arrived at the war on Vietnam soil, he stood firm on his obligation as a commissioned officer.

Two infantry companies typically protected the FB: sometimes, only one infantry company. In the FB, there would be an artillery battery, six 105 mm howitzers, and cannon cockers the same number as the infantry company for manpower. For protection from enemy fire, shielding each tube or artillery was a parapet of sandbags, and serving as overhead protection for the men's bunkers were logs and sandbags.

The FB was considered as rest and relaxation (R&R) for the infantry company, especially if there were two infantry companies. For the FO team, that was not the case. Why? Because when there were two infantry companies, it meant that after an infantry company had come back from field duty that lasted either about two weeks, three weeks, a month, five weeks, or longer, they would be replaced by the other Infantry company guarding the FB.

The FO team, on the other hand, after coming back to the FB, would again move out the following morning with the fresh unit. When they went out, the infantry would increase clearing the fields of fire to give more room for the projectiles they would fire from their artillery in their FB. They would also run quadrant sweeps, where they would go out of the perimeter two to four clicks (kilometers) and search a quadrant of a circle to look for any signs of the enemy intending to over-run the FB.

The person walking point is the person responsible for warning their fellow troops of the enemy's presence in preparation for battle, would go out with the infantry company and take the lead as they walked ahead of them.

They would climb mountains and hills all day long in search of Charlie to make contact and hoping not to. They would be observing nature for a couple of weeks, which my father said could be as simple as kneeling on one knee while looking at their map as they lit it with the red filter of their flashlight and unexpectedly observing a long snake slithering over the top of their boot while it was out hunting.

When my father assumed the FO's position for Delta Co, 3rd of the 8th INF, he realized the enormity of his responsibilities. Apart from reporting the enemy's presence to the infantry once they spotted the enemy, as mentioned earlier, the FO is in charge of directing the artillery people where exactly to fire explosives. So he needed to be an excellent map reader. He had Army cartography to thank for being able to navigate precisely. He commented that Army cartography was excellent. He would be within 15 feet of the area he was in on the map, even when the map provided by the Army had the caveat saying it was a projected guess of the actual geographical features.

In the Army, the infantry is regarded as the "Queen of Battle," while the artillery is considered the "King of Battle." My father explained that this distinction exists because, when troops engage in a firefight, have an objective to capture, or face a numerically superior enemy, it is often the artillery that causes the majority of enemy casualties.

When operating in the field, my father would carry a substantial amount of ammunition to last the entire mission—specifically, 2,000 rounds of 5.56 mm ammunition weighing about 40 pounds. In comparison, a single 105 mm artillery shell weighs approximately 37.5 pounds, nearly the same as all the ammunition he carried. If 200 such artillery shells were fired in support, the amount of firepower delivered would far exceed that of the entire infantry unit's combined small-arms fire.

This comparison illustrates the immense destructive capability of artillery compared to infantry weapons, highlighting why artillery is considered the "King of Battle."

Forward Observers (FOs) play a crucial role in battle. By monitoring the front lines, they provide the infantry with advance notice of the enemy's positions. More importantly, their instructions on where and what type of explosives to fire help inflict greater casualties on the enemy. In just 10 to 15 minutes, an FO, with the support of artillery units, can deliver as much firepower as an infantry company would in two days.

Specifically, in critical situations, an FO can effectively match and even exceed the firepower of an entire 80-man infantry company. The total firepower carried by such a company is equivalent to 80 rounds of 105 mm artillery shells. When an FO coordinates fire from multiple batteries—each firing six howitzer tubes at a single target—they can replicate the entire infantry company's firepower within just 15 minutes.

In contrast, the ammunition carried by each infantryman is intended to last for a couple of days.

Firing artillery is not just randomly throwing explosives out in the forests—that does nothing. It is firing at the enemy that counts. That requires precision and caution. The FO has to correctly read the map and relay instructions to the artillery people who will be firing the explosives. He also has to make sure not to harm nor kill the men in his unit. So, it is not just the efficiency in firing explosives that relies on the competence of an FO, but also the lives of the men in the infantry company.

It is my opinion that every 15 minutes an FO stays alive during a battle is equal to two days of infantry power. When the FO stays alive, so does the infantry company. That is how powerful an FO is.

Nevertheless, the infantry, too, has an enormous responsibility for the entire company's lives. In either position, whether infantry or FO, it would be a burden to live with the failure if they contributed to the cause of needless deaths. I'm trying to say that each position has its respective important roles. Every soldier has their life on the line.

My father gave this little trivia: a late-night TV host would earn approximately $15 million per year. That raised the question, "Would you want to earn that much in a year as an infantryman or an FO in combat?"

That was the same amount of money, but compared to a late-night TV show host's job where work is done in a studio, an infantryman or FO would be out there, risking his life.

Moreover, anyone assuming a job on the battlefield can have their life at risk. For example, those on listening post (LP) duty were considered the unit's early warning line and eyes and ears. To give a descriptive picture, if a couple of soldiers on LP duty, say a private first class (PFC) and his LP buddies, were caught off guard, they can have their throats slit by the enemy, which will then allow the enemy to ambush a company. Since there would not be any warning from the LP, the company could be overrun. Everything a soldier did or failed to do out in the field had consequences.

When my father underwent his first fire mission, he felt scared. He compared the fear he felt to how one would feel when doing particular things in life, like speaking on a radio, talking to a microphone, driving a car, etc. Especially if it will be done for the first time.

In the case of his first fire mission, it was calling in defensive fire concentrations (DEFCONs) at a known point, calling in artillery about the four quadrants 300 meters out of your nighttime position. In his first time doing it, he wrote out what he had to say to avoid screwing up.

My father characterized Charlie as tenacious. They were more on fighting face to face, as they did not have much artillery going on except for mortars, nor did they have bombers as the US Armed Forces did.

In adjusting friendly fire from a battery, 50 meters is the closest distance it is supposed to be adjusted. My father often did it within 25 meters. He claimed that firing within 25 meters of his position never caused any casualties. The artilleries he would ask to be fired from the firebase was done very professionally. The artillery people would observe the proper way of firing artillery, leveling their bubbles after every round before firing the next. When it is done right, the artillery always comes accurately. If Charlie was 25 meters away from their artillery, he had a chance of living. If he was 50 meters away, he was toast.

On the contrary, my father expressed his disappointment with mortars. "Having the Infantry fire mortars was a different story!" he commented. He had a bad experience with mortars, which then caused him to lose faith in them. One night, the incident happened when my father's unit was to finally retreat to a firebase after weeks of being out.

Although there was not enough daylight for them to make it, their proximity allowed them to avail of artillery support and infantry mortars. He had the DEFCONs all set, and things seemed fine until there came a turn of events about six hours later—their listening posts at all four quadrants, each a group of about three men, reported that they heard enemy sounds. Each of them knew how far from their position that my father had adjusted their DEFCON. They informed him it was from 50 meters in front of them. So, he adjusted fire 50 meters ahead of their position.

Keeping track of exactly where the last round came in at each of the four quadrants was a challenge for an FO when at the same time, he would contact all four quadrants simultaneously. This matter was not even taught in the Basic Artillery Course. Thankfully, with my father's quick thinking, he was able to come up with a method very quickly. The method involved these three steps: 1. Get out a notepad and put the main group in the center. 2. Mark each outpost from the center. 3. When firing for that outpost, mark the last round fired with a higher number or letter. "Problem solved," my father thought.

However, there was a shortcoming on the part of the mortar squad. My father aimed to fire an 81 mm mortar concentration for the enemy to be driven right towards it. It would've been perfect. Unfortunately, the result was not what he expected. He asked for five rounds to be fired at that concentration, but the mortar squad's execution turned out horribly. Although one round landed correctly, the following three rounds landed inside their perimeter, and the fifth fell outside their perimeter, just right below it.

Due to the mortar squad's carelessness, not level their bubbles between each round. Had they leveled their bubbles, the mortars would have been fired consistently, and they would have fallen exactly where they were supposed to land. Their malpractice could have caused casualties. Thankfully, there were none. They explained that they thought my father wanted the mortars fired quickly, so they fired the mortars hastily without bothering to level the bubbles. Because of that incident, my father never bothered using the mortars again, except for this one time when their unit was being overrun.

When he worked with the mortars again, he tried to be cautious that he had to ascertain the mortars with a call for check fire, which means to stop firing to determine whether the rounds that landed nearby were friendly fire or coming from the enemy.

After a couple of weeks out with the infantry company, the FO team would return to a firebase. They did not get to have some R&R in the firebase that the infantry got to enjoy after being out for a long time. The FO's are always on the move out in the field.

However, unlike those in the infantry, particularly the Military Occupational Specialty (MOS) people, the FOs were not eligible for the immediately recognizable Combat Infantryman's Badge (CIB). This badge is given to infantrymen who have demonstrated mastery of critical tasks and skills and have been in contact with enemy forces. Moreover, minor technicalities can also make a big difference in not receiving the CIB, such as wrong MOS.

Unfortunately, in the case of the FO, even if they fought well in battles, they were not given that kind of recognition. Looking into it, though, the FO's are more infantry than the infantry, since they were out in the field day after day. Nevertheless, the FOs at least had the solace of being an integral part of an elite unit with a special technical term reserved for them: "Bait."

Another downside to being an FO was when they would move in for the overnight stay, the FB would be overcrowded. There was no room left for them to sleep. So, they would try to make do with the available space by the parapeted 105 mm tubes. They would set up their beds inside the parapets or around the tubes to sleep for the night. The only fly in the ointment was that as the FOs tried to get a good night's sleep, there would suddenly come the loud sounds of harassment and interdiction (H&I) fire being fired off by the artillery people. Those were just rounds fired at a chosen point within the area of operation. The rounds would interrupt the FO's peaceful sleep as the artillery people fired it without warning but surely with glee.

When the infantry company and FO team went out for field duty, they would start off every three days carrying equipment almost as heavy as their own weight. The equipment my father carried was as follows:

Pounds: Equipment

6 Rifle

6 A minimum of 3 one-quart canteens (a pint is a pound)

8 2 Collapsible canteens that held 2 quarts

3 Entrenching tools

5 Web gear (pistol belt, harness, and bayonet)

6 Large Alice pack and Aluminum frame

4 Poncho and liner (light blanket)

6 3 hand grenades

6 2 smoke grenades

15 An equivalent of about 9 C-rats (meals)

40 2,000 rounds of rifle ammo (there are no ammo stores at ambush sites)

10 20 round magazines (actually, they carried about 17 or 18 round magazines. The magazine springs were not trusted to feed reliably under full compression tension for extended periods.)

4 Helmet and steel pot

7 Uniform and boots

2 Personal gear: cigarettes, razor, writing paper, SS mirror (shaving), and beer can opener

——

128 pounds approximately (Most people did not carry 2,000 rounds of rifle ammo, but they carried the same equivalent weight, maybe in place they might have had a radio battery or two or three grenades.)

For radiotelephone operators (RTOs), the ammunition was not included in the equipment they carried, unlike my father, who was an FO. However, they carried a 35-pound radio and a 6-pound battery. My father and his recon sergeant would also carry a spare battery.

The medics, too, had their equipment to carry in their medical bag along with their rucksack. In total, the weight that everyone carried was approximately 128 pounds, which had little difference to their weight.

As re-supply neared, though, the weight they carried decreased to about 24 pounds, which was 12 pounds each of water and food. —"Life was good, just a walk in the sunshine!"—as my father remarked.

When my father started his service in Vietnam in early September of 1967, he weighed 175 pounds. Later in the war, he lost a lot of weight from the strenuous field duty activities, for example, hiking for miles in jungles, up in hills and mountains. Their mission was to search for Charlie and make contact while carrying a huge, heavy rucksack and eating only small amounts of food.

After some time, he went from weighing 175 pounds down to 130 pounds. His 28-inch waist jungle fatigues had gotten loose. That was even after he had been fattened up for a couple of weeks when he was hospitalized after being wounded twice two months into his Vietnam tour on November 9, 1967.

His first wound was inflicted by a 60 mm mortar round. His other wound was inflicted by a hand grenade, which caused pneumothorax, meaning his lung collapsed, and it was also full of blood. He was then hospitalized for three weeks in-country and four weeks in Japan.

Being away from the war, he had regained some body fat, though his weight was still less than when he started his service.

For soldiers, being wounded was a possible part of their job. It is a sacrifice they made in their military service. When my father was in the hospital in Japan, he met a fellow lieutenant hospitalized there. The guy got shot when he was walking point. It was only him that got shot. He wondered why it was so. My father guessed that the shiny metal officer brass he wore when he was walking point—not embroidered nor subdued—may have made him a target of the enemy.

Among all the equipment carried by the infantry company and the forward observer (FO) team, ammunition was the heaviest. Machine gunners, for example, might carry 2,000 rounds of ammo, weighing about 40 pounds. They carried so much because there was no resupply points in the field, and if they encountered the enemy, the nearest friendly units could be 7 to 15 miles away. To support them, they relied on artillery.

My father was a First Lieutenant and Ranger-qualified

artillery officer attached to the infantry company. When they engaged the enemy, they would have reinforcement from artillery, and in short order Huey gunships and Air Force fighter planes.

They also brought a beer can opener with them, but it was barely ever used for opening drinks. "Yes, we would get a cold six-pack every night—in our dreams!" my father sarcastically remarked. In reality, the can opener was least likely to be used for beer or soda. As my father recounted, "We only had soda or beer twice the entire time I was in Vietnam."

Their can opener was more often used for fascinating themselves by fabricating a furnace with it. They would use that furnace to heat their C-rats. To create fire for their furnace, they would use a C4 block, a white plastic explosive. They would place the C4 in the furnace and burn it. They would make six punctures in the top half and four punctures in the bottom half on their C-rats. That was the closest they could get to a TV for entertaining themselves. My father terribly missed soda, that he vowed that he would always enjoy a cold soda when he got back home.

Indeed, ever since he got home, he had been bringing a soda up to his nightstand for him to enjoy every night for the past five decades.

For sleeping, they had an air mattress. It weighed five pounds, and my father said he would wake up every day with zero air in it despite putting duct tape on it. They just had to make do with it because, at that time, the Army did not have an alternative. They did not have the sleeping foam pad that the Army now has.

Sleeping out in the field was uncomfortable for them, as they would lay on bamboo stubs, rocks, roots, etc. If there came a downpour of rain, it would be another added variation of miserableness. But they quickly got used to the inconvenience.

Although it would be scorching in the jungle during the day, it would be cold at night. They each had a poncho liner, but just one poncho liner did not cut it. My father's commanding officer (CO) was fortunate to have had an Army sweater; my father and the others were envious.

There was this one cold morning that my father recalls. He was shivering from the cold and yearned for a hot drink to feel some warmth. It was the day that re-supply was due. At that time, my father had no water but only had a partial canteen of Kool-Aid.

Desperate for some coffee, he tried to make coffee out of the Kool-Aid. Unfortunately, it was undrinkable with its disgusting taste. So, he just threw it away.

As the end of their three to five-week-long field duty neared, chances were they would have no change of clothes left. The only thing they would have left were socks. After three weeks, they got a re-supply of clothing from Direct Exchange (DX). They would be provided with clothes they could wear for the next few weeks. They would try to find clothes that fit them right and were not too small.

Their socks were the only things they would try to wear for a couple of days, try to get a chance to wash out, and put on a different pair of socks.

Being away from loved ones, mail was the only way for soldiers to keep in touch with them. Every three days and along with re-supply, they would have mail call, wherein soldiers who have mail get called to take their mail. When they received mail, they would have to burn their letters after reading. Letters were not to be kept, as the enemy may use them as propaganda if a soldier got killed or captured. The enemy even had a propaganda station. When my father's unit would be listening to the radio at night, they would pick it up sometimes, playing music. Then they would be surprised to hear the enemy speak to their unit, saying things that they had found out through obtained letters like "Welcome back, Capt. Jenkins from D 3/8, how was your R&R in Australia?"

It was earlier mentioned that some troops tend to be stubborn. Some can also be quite reckless. Such manners of theirs can lead to terrible consequences—even death. "Some units were not as closely supervised as they should have been," my father remarked. Some troops took the C4 from their defensive perimeter claymore mines for heating their c-rations. It seems convenient for getting your meal nice and hot, but this practice had led to many deaths of Americans as found in battlefield analyses that had been carried out.

When the explosive is taken out of the claymore mines, the rate of the ball bearings getting blasted out would decrease from 1500 feet per second to only 100 feet per second, which results in American troops getting hit.

One night, a defecation-related accident led to a soldier's death. It was because he did not follow the correct way to crap. Supposedly, they were only to do your business as soon as their nighttime position, the spot where they would be spending the night, was selected, while it was still daylight. It was frowned upon to do it in their perimeter at nighttime. When they needed to take a crap, they had to do it near the end of the field that had been cleared for their nighttime position, where they had to dig out a foxhole and clear a field of fire. Well, the soldier did his business around almost dusk. When he came back in, he was mistaken as an enemy and shot dead. Right after that incident, another mishap happened later that evening. These two companies were on the same mission, traveling the same route. My father was in one of those companies.

An FO from another company called in a fire mission in response to the listening post reporting activity. He had done it carelessly without checking how the shells' angle would pass overhead as they fired from the FB's battery that he requested fire from. It would've caused tree bursts to go off overhead if the FO's requested rounds were honored. It was a good thing my father was quick to call in check fire on him, which canceled the fire mission, thus preventing the tree burst from happening. Bell, who was the two companies' senior, saw that my father's intuition was right and backed my father up. That resolved the issue of the shells being fired.

The following morning, a senior artillery officer, a major, flew out via an artillery chopper. He shot an azimuth with his compass to check the area and took off—that confirmed that my father was right. If the shells passed overhead and hit the trees, they would have exploded and injured or killed the soldiers in my father's unit.

Each unit had control of firing in its area. Another unit is not allowed to fire into another unit's area. The other unit wanted to fire into my father's unit area. Anything that a soldier does or fails to do in the field has its consequences.

Then there was this one very cold morning they woke up in this recently bombed area with deep bomb craters that they had moved into. They were given a mini R&R that day and were allowed to stay in for an extra day, for the day before when they had contact with Charlie. That day, they held a little ceremony for a few men that had been killed.

The previous evening, they received a resupply of food and ammo. To keep warm amidst the cool temperature that morning, someone thought of starting a fire in a crater. He made use of the C-ration cardboards and then he tossed in the empty ammo crates. My father said the warmth felt good, but he moved away for some reason.

Luckily, when he moved away, he was spared from an explosion that went off when someone accidentally tossed in a crate with a grenade still left in it. The grenade cooked off and wounded several men.

My father, too, had one of those reckless moments. It happened when they were blowing up trees to make room for an LZ in a resupply clearing area where a chopper can lower supplies to them. Naturally, they would protect themselves from the explosions with anything they can shield themselves with. My daring father, who was behind a log, suddenly rose up when the last "Fire in the hole" warning was issued just so he could take a picture of the event. Gigantic pieces of wood and splinters the size of railroad spikes flew at him.

He said he almost exited the gene pool at that point and did not even know how he survived. Regarding situations as such, he expressed, "Stupidity is not inherited. Young men in battle die before passing on that gene."

As they walked miles and miles every day, the men experienced all kinds of terrains. One time they came across this terrain with boulders around 20 to 30 feet in diameter scattered all over the place. My father was in awe of the view in the area. There, he mused, "Little did I know I would make my way home in West Rockhill Township."

The terrain had stuff that looked like chimneys like in West Rockhill Township. At that time, he was still a newbie. The area reminded him of home. My father remembers taking a break there. He took a few moments of rest as he plopped down on his rear, leaned against one of these pillars with his rucksack and had a smoke. When he had smoked half of his cigarette, while he was enjoying the view, he engraved in his mind fire ant chimneys and on-looker's amusement towards them.

Other times, they would find themselves in tall grasslands, as tall as six feet. There were times when they would be in an area in the forest that would overlook a sea of grass fields. Occasionally, they would notice Charlie out there. They would spot Charlie's flashlights as he moved about.

Once, when they caught sight of Charlie from afar, my father borrowed Bell's starlight scope to get a better look at them. Preparing to attack the enemy, my father asked Bell to hand him an M-1 D rifle. The starlight scope could not be mounted to the rifle though. The M-1 D rifle was the sniper version of the M-1 rifle, which was more suitable for shooting at night.

My father was an expert in shooting. That is all thanks to all the practice he got back when he was in the Reserve Officers' Training Corps (ROTC) in college. There, he took advantage of the available army rifle range and fancy .22 caliber rifles. He would shoot 50 to 100 rounds between classes every day. Then in his senior year, during the ROTC six-week training, he was awarded second-best shot out of a thousand other students.

Unfortunately, when they saw Charlie that night, he did not get to shoot at them. They did not have the right rifle for that moment. The rifle was longer and heavier, and its bullets weighed three to four times more than that of a .223. It would have just been an albatross about my father's neck. So, they just moved out of the terrain.

There were also times when they were surrounded by dense bamboo. They've also been in defoliated areas, only high timber growth but almost zilch in foliage growth. Sometimes, they would pass B52 areas —areas bombed by jets called B52 bombers—cratered and just smelling of death as depicted by my father.

Then, there was also the traditional triple canopy jungle. My father even has some pictures of a triple canopy jungle with a 15-foot view. Five men were in the picture, which can be a challenge to spot. One would be lucky if able to spot three.

In October of 1967, he spent a brutal night calling in artillery strikes without rest. The North Vietnamese Army had sent groups of fifteen soldiers to test their defenses. Based on reports from the listening posts, the brave men were fifty meters out from the perimeter, McNulty estimated they faced small squads trying to sneak in and probe their lines.

He fired through the night. After each round landed, he heard "Charlie" moaning in the distance. Soon followed eerie sounds of enemy troops creeping through the dark, dragging away their dead and wounded. The jungle echoed with bugles, horns, and the chaotic noise of war. But McNulty and his team held firm. They didn't just defend their position—they fought fiercely. He figured he had killed or wounded between 28 and 35 enemy soldiers that night alone.

It was exhausting. At one point, he was managing three artillery missions at once, adjusting fire in different directions almost automatically. He had barely three hours of sleep over the past couple of days. But the following night, they let him rest. Perhaps that fierce defense made the enemy reconsider messing with them again.

Setting up camp afterward felt peaceful compared to what they'd just endured. McNulty wanted to move toward the area where the fighting had been heaviest, but his commanding officer decided he was too valuable to risk on the front lines.

Yet, they heard they might go on another mission. To the north, about ten kilometers away, a 200-man Viet Cong unit had been spotted.

Later, they experienced another attack—a small probing strike or perhaps just a chance meeting between enemy forces. Seventy-three artillery rounds were fired that night, along with continuous rifle and machine gun fire. When the shooting started; McNulty was shaving. He finished quickly, then went to work firing artillery duties as though nothing unusual was happening.

Eventually, the planned move north was canceled, probably due to gunships roaring in and firing rockets. It was loud and chaotic, but McNulty remained calm, collected, and accurate.

My father described being in the field as almost a Biblical experience. They often faced probing attacks from Charlie, who sent small groups to test firebases and check their defenses. If a firebase appeared weak, larger enemy forces would try to overrun it. My father compared it to the Biblical Battle of Jericho, where attackers blew horns and bugles, unsettling their enemies effectively. As he said, "It was so primitive it was unnerving—hearing dozens of bugles sounding all around you."

Early in the RVN War, there was a strange belief that the enemy could detect scents. Some infantry officers banned aftershave and cigarettes on patrol because of this folklore. My father thought this was unrealistic. Soldiers wore the same clothes for weeks and went without bathing for long periods. He explained that detecting the enemy went both ways—when soldiers smelled the enemy; it wasn't aftershave but weeks of accumulated sweat and grime. Apart from a quick stream bath near the end of the monsoon season, my father couldn't remember having a proper shower for almost two months.

My father occasionally flew in helicopters. From above, he saw that Vietnam's terrain looked shockingly like the moon's cratered surface. The heavy bombing ruined teak forests, making them unusable for generations due to shrapnel.

At night, my father remembered hearing odd sounds

resembling the phrase "Fuck you." These calls came from what soldiers named the "Fuck You Lizards," creatures that often cried out together, creating a strange chorus.

These lizards were annoying, but an even greater nuisance were the leeches crawling into soldiers' clothes. Ironically, these pests became a form of entertainment. Unlike base camps and firebases where recreational activities were available, there was little entertainment out in the field. Base camps had clay basketball courts, and firebases sometimes had softball fields or gambling to pass the time.

In contrast, entertainment was rare in the field, making leeches an unlikely amusement. After tiring days trekking through forests and mountains, soldiers relaxed in their bunkers at night by counting leeches on their bodies and deciding how to remove them: burning them with cigarettes, using DEET or simply pulling them off.

During their first major battle in Dak To, near the tri-border area, my father's unit learned a costly lesson. They were a new unit that replaced the original infantry who had arrived by boat in 1966; his unit came by airplane in 1967. My father arrived a month after the transition, so he was unaware if the previous unit had shared any lessons. In this initial engagement, they discovered that the enemy heavily relied on snipers positioned high up in the trees.

This tactic was different from World War II scenarios where Japanese soldiers tied themselves to palm trees—Vietnam had no palm trees but dense triple canopy jungles. A triple canopy jungle is akin to a foggy sea. It consists of rich vegetation growing in three distinct layers, reaching tremendous heights from 10 to over 100 feet. The first layer includes bamboo, shrubs, and vines up to 10 feet high. The second layer comprises young trees ranging from 10 to 40 feet tall. The third layer features mature trees soaring 100 feet and higher. This dense vegetation severely limited their visibility as they moved through the jungle; they could barely see anything around them.

However, a sniper positioned high in a tree could easily see through the lower canopies. This gave the North Vietnamese Army (NVA) a significant advantage. From the ground, soldiers couldn't determine where sniper fire was coming from. It's similar to standing on a foggy shoreline: when you're on land, you can't see a boat out at sea due to the fog. Conversely, from the boat, you can see people onshore through the fog. Likewise, in a triple canopy jungle, the enemy above could see them clearly, while they couldn't see the enemy.

Battling in such a jungle meant fighting unseen enemies. From the front, they might face fire from a .51-caliber machine gun but couldn't pinpoint its exact location. Additionally, isolated sniper shots would come from the treetops. They were being attacked from multiple directions but had to focus primarily on the frontal assault while hoping not to get shot from above. Moreover, there were other threats like bunkers with small arms, light machine guns opening fire, hand grenades, B40 rockets, and mortars.

According to my father, *Saving Private Ryan* was the first war movie that came very close to duplicating the chaos and terrifying events of landing on foreign soil. When it comes to what an actual battle sounds like, my father said that the closest version he could offer to a civilian who would like to replicate a battle experience is to go to a gun range with eight of your buddies who own semi-auto AK47 and M16 rifles and bring a good Sears 35-gallon galvanized trash can. Position yourself at the 33-yard mark and stick your head about 1.5 feet into the trash can. Then have your buddies put the muzzle of each of their rifles in the trash can and open fire continuously as fast as they can fire for 15 minute periods, followed by quiet periods of 2 seconds to a minute for the next few hours.

In my father's assignment as an FO, he got to work with all of the companies. As he worked with different companies, he got to work with different captains; some, my father commended as competent and well respected, while some, he considered unfit to be captain.

One captain that my father found as an extremely rare individual was Capt. Falcone. According to my father, the troops he had handled loved and admired him. He regarded Falcone as one of less than five individuals that he had met in his life who was so well-liked and competent that anyone would do anything for him, and he, in turn, would do anything possible for you.

Another captain my father commended was Capt. Taylor. Both captains were West Pointers. He had encountered these competent captains when he worked around the west quadrant. They had both been to Ranger School and knew their stuff. Nobody would have ever guessed that both would, unfortunately, be dead in a month. Then there is also Bell, who was also a Ranger School graduate. He was a competent captain and well respected.

But there was one captain, Capt. Owens, whom my father deemed unfit to even lead a pack of Cub Scouts. My father became an FO for him once. When they were out for two to three weeks, they missed food and water re-supply for some reason. So, the next day until evening, they had to just make do with whatever they had, which was close to zilch. My father recalled having breakfast like in Ranger School: three or four packets of salt and pepper.

When midday came, a point radioed back that he had come across a stream. When the troops reached the stream, they dropped their pack and rifle to fill their canteens with water, thereafter, drank from it immediately without even placing Halazone tablets, which were for purifying the water. Supposedly, one or two Halazone tablets were to be dropped into the water, and they had to wait 20 minutes for the water to be purified before drinking.

Not only did Owen's men not have their weapons and

gear with them, but he also had not sent security out for a lookout. So, my father went to Owens and started yelling at him. He told him to get security out on the far side of the stream as well as up and downstream and to tell his men to get their weapons and gear. He even told Owens that he was not fit to lead a pack of Cub Scouts. After what my father had done, he thought he was going to get court-martialed. Fortunately, he was not. When Owens finally got security posted, the security posted upstream reported he found two dead NVAs in the stream. The reaction of the troops drinking the water without dissolving Halazone tablets when the word about the dead NVAs up the stream reached them was almost comical. Out of disgust, they started to swallow the tablets whole.

As tenacious fighters, rumor has it that the NVA took dope to get hyped up and do Banzai-type attacks. Just one con that they had was that they had no aircraft nor real artillery support. The outcome would have been different if they did. The only weapons they had for reinforcement were mortars. On that matter, my father remarked, "Poor Charlie. He came south, probably pushing a bicycle full of supplies, knowing he would be away for years. No R&R to look forward to and forced to fight against B52s, artillery, Puff the Magic Dragon, air and artillery superiority, and no air re-supply."

Nevertheless, my father commended Charlie, saying they did a fabulous job under very adverse conditions. It was usually North Vietnamese regulars that their unit chanced upon. They did not encounter any villages nor Viet Cong.

The NVA had established their territories in the jungles of Vietnam and had sheltered themselves in the forests for 20 years, typically beneath hills, where they had dug tunnels. One time, my father's unit went up the Chu Goungot Mountain. When they reached the top of the mountain, they found a stairway made with logs that were six-foot-wide stacked as steps going all the way down the other side. It even had handrails. Another time they found a vertical shaft that was about six feet square. Out of curiosity about what it may have been for, they dug it out. After digging about six feet below, they found logs laid flat in a row. Then after digging out another six feet, they again found another row of logs perpendicular to the first. Every after six feet of digging, they would find a row of logs. It was like there was no end. Until they finally decided to just blow up the area with a C4. They eventually just gave up, as about 5 or 10 minutes later, they noticed a plume of smoke or dust rising above the trees from approximately a quarter of a mile away. They just moved out and did not get to find out what it was.

In war, when soldiers make contact with the enemy, anyone could get wounded or killed. So, soldiers serving in the war would try not to get too personal with their fellow soldiers. They would simply know their fellow soldiers by their first name, call sign, or 2-4 element, 4-5 element, or 4-6 element code name when they talked on the radiotelephone. That is because when someone close to them becomes a casualty, it can affect them emotionally. Not getting too attached to anyone makes it easier for them when people were getting wounded or killed.

With the rush of adrenaline and the invincibility he had somehow acquired after already being in the war for some time, he volunteered to go out to look for the sniper along with half a dozen other men. He loaded up on grenades, about four to five of them, which he had hung from his pockets and took about 20 spare magazines and went off with the rest to find and fight off the sniper. No rucksack this time, since they were only going a couple of hundred yards out of their perimeter.

Out there, he had this feeling that eyes were on him and that he was dead on someone's sites. Thankfully, because of my father's smart thinking—of carrying grenades around his pockets—the sniper did not pull the trigger. If the sniper did, then my father would've gotten him with the grenades also, as the grenades would have gone off, blasting toward the snipper. However, they did not find the sniper. They came real close though.

There were several other times when my father felt eyes on him and that he was dead in the enemy's sites, like when he would move outside of the perimeter as he fired DEFCONs. Everybody in the FO party would feel those eyes. Then they would all quickly move back into the perimeter.

Fast forward to 30 years later, my father's sixth sense

that he developed at war came in handy when he traveled to Ireland. During the time of his travel, there was a conflict between Northern and Southern Ireland. At that time, the Irish Revolutionary Army (IRA) was known to blast bombs to target British civilians. He was with my mother in a town square in Northern Ireland, and he felt eyes on him.

Feeling that something was off, my father decided that he and my mother should leave immediately. Then just as my father had suspected, half an hour later after they left, a bomb went off. Six died from the bombing.

CHAPTER SEVEN

The Battle of Dak To: Sacrifice, Survival & Strategy

It was early in November 1967 when a series of battles struck in Dak To, wherein some had turned out to be the toughest and bloodiest battles in the Republic of Vietnam (RVN) War. It was among the North Vietnam Army's (NVA) series of offensive initiatives along with the battles in L⊠c Ninh, Song Be, and at Con Thien and Khe Sanh, which were known as the "Border Battles". The contact between the NVA and American forces in this battle was considered major engagements in the war. At that time, the Battle of Dak To was the largest in the Vietnam War. Although it was in the last tat, it was a general battle. The engagements in Dak To lasted from November 3 to 23. It all began when the American forces had gotten word from a defector about the locations where the NVA had been staying.

Upon having an idea where NVA was, the next thing for the American forces to do was to catch them to make contact. So the American forces launched Operation MacArthur in the Central Highlands at Kontum Province. The army brass—high-ranking officers—relied on a specialized unit for that specific task. They ordered this unit to search for two or three NVA regiments.

Although they knew where Charlie was, what they did not know at that time was that the NVA regiments were working on two goals: 1. To find and utterly destroy an American unit, which was them, for a public relations story in the newspapers and 2. To move out of Laos and Cambodia and go down to the coast for the Tet offensive.

The Tet Offensive was a series of attacks the NVA planned to carry out in more than a hundred of South Vietnam's cities and outposts during Vietnam's Tet holiday, a grand, sacred, and most important holiday for Vietnam, which celebrates Vietnam's Lunar New Year with all sorts of festivities that showcase Vietnam's culture. Before the Tet Offensive, what usually happened during the Tet holiday was the opposing sides would have an armistice. That was similar to what happened during Christmas day back in World War I, the Germans would not shoot at the US troops on that day and vice versa. Then after that holiday, they returned to business as usual. That was not the case on January 31, when the Tet Offensive commenced.

Together, the Alpha and Delta companies of the 3rd Battalion 8th infantry teamed up in searching for enemy regiments. Only two companies were needed since the task was to find only two regiments.

They were sent off to Dak To to do a helicopter combat assault (CA). "Lucky us. Scared shitless! Such great odds!" my father remarked sarcastically on the task. In a helicopter combat assault, what happens is that about six guys would get in the chopper, and when they've reached their destination, they would jump out of the hovering chopper, move out, and then fan out. Then there would be other helicopters following that carried other troops to the destination and so on until all of the infantry companies were brought to the location. There were about a couple of hundred troops who landed.

Men in war do not know the luxury of distinguishing between the days of the week. They might know a date a battle happened because it was very memorable to them. On November 4, as far as my father can remember, troops from the Alpha Company came. As they arrived, they did the usual—they dug out fox holes and chopped down trees that were about five to six inches in diameter with their machetes for overhead cover.

The following day, November 5, some of the troops from the Alpha Company ran into gooks when a commanding officer (CO) sent them out. They got into a firefight and a lieutenant got killed. Capt. Taylor of the Alpha company along with another element came to reinforce them and recover the dead. Unfortunately, the captain, too, got killed.

However, they were able to capture the gook who killed the lieutenant. The lieutenants of the Alpha and Delta companies thought at that time that it would be reasonable to send the gook's SKS rifle to his widow for her to hang it over her fireplace. Being in combat does tend to affect one's thinking.

On November 6, the first thing they did, as usual, was their routine task of destroying their bunkers and overhead covers before going on the move. It was part of their routine to destroy their bunkers before they left so that the NVA would not benefit from the bunkers that they had made and left behind. If they had left their bunkers as they were, then the NVA could come across and use those ready-made bunkers, and they did not want that. They would dismantle and throw away their overhead covers and as much as possible, hoped the used overhead covers would fall down the highland.

That day, my father felt mischievous and left behind for the NVA a trophy (crap) in the bunker dirt. After having destroyed their bunkers, they were off to their objective: Hill 724, going downhill and through the bottom of a valley to reach the hilltop. On their way to Hill 724, the Alpha Company took the lead. In their hike up the hill, the Alpha Company found booby traps made of tripwires and hand grenades. As Delta Company was in the rear, they were given a heads up by Alpha Company on the booby traps. On top of that, there came an airburst of an artillery round over Delta Company's position. Curious about the artillery round, immediately, my father decided to call in a check fire. Despite calling check fire, my father was still baffled, as it was not him firing at all, yet all the firing in their area had ceased. He wondered that perhaps it could have been H&I being fired by someone outside their area of responsibility, or maybe it was friendly fire, or maybe it was Charlie.

The incident caused Delta Company two casualties: one killed in action (KIA): Sgt. Greenwood and one wounded in action (WIA): Walter Gross who lost a leg. Out of panic, they returned to their caved in positions and re-dug their foxholes. To my father's surprise, he hit the trophy he left behind when he re-dug his fox hole. He had to clean his shovel well after that.

Then on November 7, after their routine task of taking down their bunkers and overhead covers, off they were again to their main objective with the same position as last time—Alpha Company in front, followed by Delta Company. So, Delta Company's FOs were in the mid-point. On that very day of November 7, they met several warning signs that it was going to be a terrible day. There had been more booby traps reported that day.

Then came a big surprise for them by the time they were several hundred meters away from their nighttime position and five meters towards a muddy stream. They were already nearing their objective, but before they could even start their ascent, all hell broke loose. They had been ambushed with a 51-caliber heavy machine gun. They got caught off-guard, as they had not seen it coming and had no idea where the machine gun was firing from, whether it is positioned about a half of their way, two-thirds up the hill, or at the hilltop.

They were in a dense triple canopy jungle with rich vegetation they could barely see-through. They were all terrified by the machine gun that in an instant, they all dropped to the ground. Some of them were lucky enough to have something to take cover with, like the one-inch diameter bamboo or a fallen log.

However, others unfortunately had nothing. My father was able to take cover behind the bamboo. Caught up in heavy action, my father called in fire on the 51-caliber heavy machine gun. He felt great fear, and probably everyone felt the same. Most likely everyone there, Capt. Bell, the platoon leaders, platoon sergeants, squad leaders, and everyone below them felt a sickening feeling in their stomachs.

My father guessed the 51-caliber heavy machine gun may have been the root cause of it but what gave them that feeling in the stomach would be the flight or fight syndrome, which as explained by my father, "Causes the blood to drain from the extremities to your trunk body, which prevents the body from bleeding out fast."

Following the machine gun a few minutes later came 60 mm mortars being fired towards them, which added to their fear. The NVA put them between a rock and a hard place. Making matters even worse was they were also being shot at by snipers up in the trees, which they were not aware of.

The situation called for artillery support and my father was the one to do it. As an FO, it was one of his responsibilities to call in fire. He had done it before, and he had no problem in killing NVA troops—but only when necessary.

He did not kill any NVA when not needed. He was there to protect his men. Usually, those incidents when he called in fire at the NVA were one-sided or fortuitous events for their side. One such event was that time in Dak To when they were in a Free Fire Zone (FFZ), which is an area with no restrictions on firing arms or explosives. He was positioned on a hill, where he could see over the trees and the valley. He had been firing his first DEFCON for the night in the valley.

When he called in smoke to ascertain the precision of his map reading before firing a high explosive (HE), along with the smoke round that came in and popped, there was also a green smoke grenade that popped in the valley. Curious who could have popped the smoke grenade, my father immediately checked with Bell whether there were friendly units around their area of operations (AO). Bell confirmed there were none. He also asked the Fire Direction Center (FDC). They, too, confirmed that no friendlies were around.

It turned out that the smoke grenade had been popped by Charlie to confuse and confound them. He was trying to buy time to get out of the area fast before any explosives were fired. His plan to escape backfired, as my father was quick to verify with the captain and the FDC and had gotten two confirmations within a minute. So, my father had the area where the smoke grenade had been popped plastered. It was a fortuitous event for them.

On the other hand, it was sheer terror for Charlie. What my father had the artillery do was pump out half a dozen battery 1s, which were each 37.5-pound HE projectiles, totaling 225 pounds of explosives. Those explosives have been walked around, meaning they have moved the impact area half a dozen times by 50 meters each or so. So, all in all, 1,350 pounds of HE was fired around the area Charlie was positioned, which was roughly 200 x 200 m. All of that was equal to about 400 or so grenades going off in an area. That was accomplished in just two minutes.

But on that day, November 7, the tables had been turned. At that time, they were the ones on the receiving end. My father was a casualty himself. He was wounded by a 60 mm mortar round. Though my father described that it was not so serious, the Timex watch on his wrist got blown off. "In that case," he jokingly remarked in relation to the Timex slogan, "the Timex took a licking, but did not keep on ticking."

Despite being wounded, he still called in fire, which was a challenge not only because he was wounded, but mostly because he had no idea where the machine gun and the mortars were firing from. He had to figure it out himself. There were many possibilities on where they could have been firing.

The mortars became a high-priority threat, as many, including him, had been wounded and killed. So, he felt he needed to deduce and act on the mortars immediately. One possibility that my father thought was that the NVA may have made clearings for firing the 60s, which meant that they could have been firing the mortars from anywhere. If that was the case, then there were unlimited points on the map for him to fire at. That would have required unlimited ammunition and all the time in the world to fire away. That would have been a taxing process.

Suddenly, my father made a smart guess that it was most likely that the NVA have never made any clearings because they never would have foreseen an American unit coming into their backyard.

Also, if the NVA had indeed made clearings, then Alpha and Delta companies would have heard them taking down the trees when they came to the area a few days ago. So, my father deduced that after they caved in their bunkers and vacated their nighttime position to go off to their objective that was Hill 724, the NVA may have moved into their nighttime position and set up their mortars. Why would that be?

My father figured that if Charlie had known of their presence around their area, he may want to move into the area they had just vacated because of two reasons. The first reason why my father thought that would be was that

Charlie might want to evaluate the competence of his enemy. He would move in his enemy's previous position to observe whether they were an easy mark or competent.

As my father explained, there were some factors to be considered… like when they have selected a spot to occupy, did they merely sleep on the ground or did they dig foxholes? If the former, that would be a sign that they were an easy mark. If the latter, then they would be worth ratcheting up respect for. They would deserve more respect if they dug bunkers, felled trees, and put overhead protection from grenades and mortars.

Even more so, if it was observed that they cut grasses, saplings, and bushes with their machetes to clear good fields of fire, so they would not have to low crawl around the jungle. As much as possible, Delta Company, headed by Bell, always did the full scope. Although sometimes, circumstances would not allow them to do everything.

When they would find themselves in an area devoid of trees, so it was not always possible to do everything, as they could not make overhead covers nor would not have to cut down trees.

The second reason why my father thought that, was that since the trees were already cut down in their previously occupied area, the NVA would have taken advantage of that. The area had already been cleared of trees, so the NVA would be able to fire their mortars without the obstruction of trees that could be hit and cause tree bursts over them.

By his deduction, he eliminated the other million possible places where the NVA could have been firing mortars. So, my father shifted the artillery fire towards their nighttime position where he thought the mortars were being fired. He gave the FDC the coordinates and directed to them, "Fire on the center point and walk it around at all the four coordinates—west, east, north, south in 50 meter increments." Indeed, the mortars had stopped, but he did not know why. Nevertheless, he was grateful they finally did stop.

That time, he did not dwell on the mortars too much. It was only after that incident that he wondered about the artillery fire he called in towards the area where he assumed the mortars were firing from. For almost five decades, he was curious about that call he made. He would muse, "Did the mortars stop for lack of ammo, or what?" Another thing that had once crossed his mind lately was about the NVA that moved into their nighttime position.

He put himself in the shoes of a ranking NVA officer. He explained, "If I was a ranking NVA officer, I would have sent a couple of my best scouts over to observe what the Yanks were doing, just let them get within 50 meters and observe." His point was the NVAs were well-aware that the Yanks were out to find them.

So, the question was "How competent were the Yanks in carrying out operations?" The NVA may want to find out if they were well dug in, if they made overhead covers, and if they cleared fields of fire. The NVA can just find out, themselves, in the area Yanks have occupied rather than wait till they left the hilltop.

The NVA just had to be quick to follow them and then attack them the moment they moved toward their objective. It all seemed well-planned. In addition, the circumstances seemed to be in the NVA's favor. The NVA already had the upper hand in the superiority in number. At that time the NVA had two regiments with 5,000 men. The Yanks, on the other hand, had two companies with only 200 men. The odds were 25 to 1!

Another thing the NVA took advantage of was their knowledge on how the Yanks would be out and about—the Yanks would pull out and make their way back to their previous positions—and they would not stay in an area unless they had put up a firebase. So, the NVA took the opportunity to move in mortar crews and one or two regiments into their previous positions to set up mortars when the Yanks pulled out.

The NVA were able to benefit from their old high ground, fields of fire, and clearing. Once the NVA had set up in the area, it could have been easy to catch in the open and annihilate the Yanks because the NVA also occupied every hill ahead of the Yanks, which would have left them trapped.

Had it all worked out; they would have wiped out the two US infantry companies. It would have given them a great military victory at the battle. It also would have made great press for the antiwar protesters. However, it all backfired with my father's quick thinking.

When the NVA at the hill had spotted and attacked them with the 51-caliber heavy machine gun from their front as they neared the hill, the NVA had the idea that they had already left their nighttime position and were off to a new hilltop. That was the signal to fire the mortars from behind. The NVA had them trapped—for a few minutes at least.

Luckily, the NVA's plan was destroyed, as my father had figured out what the NVA was up to and counter-attacked the mortar crews and the company or two from behind by wiping them out with artillery.

That had altered the course of the battle. Sending more NVAs into the area would not have been a good option because they would just fire at the area again. Defeat had been snatched from the jaws of victory.

In March 2017, my father and his friend, Captain Bell of Delta Company, 3rd Battalion, 8th Infantry, visited Bell's Radio Telephone Operator (RTO) Anderson at the VA Hospital in Florida. Anderson had been hospitalized due to lung hardening, which made breathing extremely difficult, and the cause remained unknown.

They stayed for several days, and during their time there, conversation naturally drifted back to Vietnam. While visiting, it wasn't uncommon for others to discover that they had served in the Vietnam War. When asked,

"How long ago was that?" the typical reply from a combat soldier (as opposed to a REMF) was often, "Last night."

Nearly five decades later, my father learned that Mike and Bell had been positioned at the rear of their company when they advanced toward Hill 724 on November 7, 1967. Mike recounted that after moving several hundred yards from their nighttime position, they suddenly heard the loud firing of mortars coming from the area they had just left. My father, using his keen deductive reasoning, correctly concluded that they had been ambushed by explosives from unseen positions. His assumption turned out to be right: the NVA had occupied their previous nighttime position, sandwiching them between a 51-caliber heavy machine gun at the front and 60 mm mortars from behind.

Following his guidance, they managed to neutralize the NVA's mortar crews and likely the best scouts from two NVA regiments. They fired approximately thirty to fifty rounds of high explosive at a target area of roughly 200 by 200 meters, which proved devastating for the NVA.

After my father had dealt with the mortars, he then had his full attention on the machine gun. He reverted to firing at the machine gun's unseen position. Due to all the loud noise in the battle, which can be likened to putting your head in the trash can while rifles fire at the trash can, as mentioned earlier, amidst all that, my father had to yell into the radio phone as he directed fire toward the machine gun so that he can be heard and understood. My father also compared the noise in the battle to the noise in a party—the yacking away of many people—that would every now and then go silent for a second or two.

Then somehow all the attention would be on a soldier as he had just uttered an expletive timing in those few seconds of silence. "It was my time, by fate, to be the loud-mouth when the battle noise stops for a second. It was I who became the focus of attention for the NVA—party poopers," my father pointed out.

It was again another one of those things that had not been mentioned in the Field Artillery Officers Basic Course. About four or five of the NVA heard my father as he yelled into the radio phone. Since he had a radio phone, the NVA deduced that he was one of the important people that must be taken out. My father guessed their idea of a place to go to was not an Irish pub.

Being busy with his sole important duty that was talking into the radio phone to direct the FDC, my father was unaware that the NVA honed in on his voice and burst into them. Thankfully, the infantry guys were quick to take defense and killed the NVA within five feet of my father's position. He had no idea that it all even happened. My father pointed out that battles are very funny. As he had explained, happenings just even as few as two or three feet away from a soldier would be unbeknownst to him.

Even if five people had been killed and fell on the ground four feet away, he would not have known. It tends to be that way in a battle. One person's experience during a battle can even be different from the experience of another person who is just one foot away. In fact, my father had only found out from some of the infantry guys a little later that there was an NVA that died just one foot away from his side.

However, although the NVA a few feet from my father may have been killed, my father, too, had suffered from the incident. A US grenade had landed near him. Whether it was from the NVA or one of their own men, he did not know. It is because in battle, troops, whether NVA or American, would scavenge any weapons left behind or laying around and use them. So, it is either the NVA threw a US grenade that they may have found, or a grenade was thrown by one of their men toward the enemy that may have bounced back to him.

As the grenade went off, pieces of it punctured his lungs, which caused his lungs to bleed out and gave him this medical condition called pneumothorax. My father knew what hit him was a US grenade because when his front chest walls were X-rayed when he was hospitalized in Japan after that battle, he noticed serrated wire fragments. He explained that U.S. grenades have partly serrated wire coiled under tension, which fragments at the partial serrations when the grenade goes off.

Planes were brought in a little later to drop bombs in the area of the machine gun in front of them. The artillery people then informed my father that they would be under cease-fire since planes and artillery rounds would make a terrible combination. So, my father then discontinued artillery firing.

Although terribly wounded, my father somehow still managed to shift to an infantryman firing towards the machine gun. By then they were an hour or two into the battle, and it was time for deduction again. It was not until my father watched *Saving Private Ryan* that he had an idea of the term "covering fire."

Also, he used to think that in a firefight, firing would be toward a seen enemy. Well, he thought wrong. My father compared the situation that they were into a Japanese folding fan. He explained it this way: "The position of the machine gun would be the start of the fan. The bullets from the machine gun would hit anywhere from its origin to the furthest edge of the fan. A slight change in the bullet point of origin covers a very large killing coverage area, at the furthest edge, in front of it. The further a soldier was from the origin of the bullet, the greater was his chance of being hit."

That means in shorter distances, the bullet arc of coverage was smaller. If a soldier was positioned closer to the machine gun, he can laterally move just a few feet to safety. However, the farther the distance, the bullet arc of coverage was larger. So, if a soldier was at the fan's far edge, he is going to need to move 50 or 100 feet laterally to escape its field of fire and be spared from death.

Since my father already assumed the role of an infantryman at that time, he directed the troops he was with to fire halfway up and higher towards the noise of the unseen machine gun in the hill. He figured that by the time the machine gun had opened up on them, the men close to the muzzle had most likely moved away from the area for good old self-preservation. He just hoped that he guessed right.

For around thirty-something years, he worried that they might have killed some of their own men from behind by friendly fire. His perturbation subsided when it was confirmed by Bell and others at 3rd Battalion 8th Infantry's reunions that none of their men got shot in the back during that battle.

The planes for air coverage took some time to arrive. During their long wait for the bombers to reach the area, my father got to fire 20 magazines of ammo up the hill. That was all he could do because his lung was full of blood, which caused him pain that limited his movement. He could barely reload his magazines.

When the planes finally got overhead, my father put out his 38-caliber flare gun, which was about the size of a fountain pen, and shot it upwards to the sky to indicate to the bombers where their position was.

From all those World War II movies my father had seen, he knew all too well that he was worse off with the condition he was in. The pieces of grenade that punctured him got him coughing and spitting up blood. At that moment, he felt like he was going to die. Though it was funny that he tried not to get blood on his fatigues because he did not want his uniform to get bloody for his forthcoming wake.

As what my father said before, war changes your thinking. It seemed quite idiotic worrying about blood getting on his fatigues that he had worn for the past two to three weeks. Because the Army would have the dead cleaned up and clothed in a dress uniform. As he thought that time would be the last few minutes of his life, he had already written a note to his wife expressing his love for her and their son and placed it inside his wallet. As he is a Roman Catholic, he also prayed the Act of Contrition.

A medic named Tim Wilson soon came and tended to my father who seemed like he was about to pass away. Wilson tried to revive my father with mouth-to-mouth resuscitation. Then he bandaged up the holes in my father's back, probably with some compress and duct tape. At some point, my father drifted out.

He described his experience as a feeling of total serenity while the gunfire and explosions were going bye-bye happily, and he saw the white light and felt total peace overcome him. Then suddenly, he was brought back to consciousness as the medic once again gave him mouth-to-mouth resuscitation.

After about four hours into the battle, the firing became intermittent until the machine gun was completely silenced. Bell walked toward the machine gun knoll. Then my father noticed that one of the lieutenants, Lt. Johnson, was moving through.

My father did not know how he managed to, but he yelled at Johnson, telling him to stop and have his men police up the KIAs and WIAs.

Johnson refused, saying that he could not do it, and he had to get to the top. Irked at the lieutenant, my father pointed his rifle at Johnson and threatened him that if they did not take the KIAs and WIAs and took a step forward, he would shoot Johnson. Out of fright, Johnson then decided to do as my father ordered. If he had not reconsidered, he would've been dead.

On the other hand, it was good that the non-commissioned officers (NCOs) did as they were supposed to with the rifles of the killed in action (KIA) and wounded in action (WIA) without being told. Earlier it was mentioned that troops of either side would scavenge any weapons they stumbled upon. So as usual, the NCOs destroyed the KIA and WIA's rifles, so they would no longer be usable. The NCOs took out the rifles' bolts and buried them in mud by the stream. Afterward, they also beat the rifles on anything they could pound them on.

On their way to the hill, they had to cross a mud area and stream that was one foot deep. My father was too weak to cross them, so some guys made a poncho litter for carrying him across. He also asked his fellow soldiers to get the remaining 1600 rounds of ammo from his rucksack and also search for more ammo in the other casualties' rucks. Afterward, they then moved towards the hill. My father had been carried in a poncho litter by four men as they crossed the stream.

In going up the hill, he did not want to remain tied up to those four men, as he did not want to be an added burden to them. So, he just asked them to let him hold onto their web harness and help pull him up the hill.

As they went up the hill, my father spotted a medic's rucksack left open, exposing all sorts of compresses and other medical equipment that it stored. He could not fathom why they found a medic's rucksack laying around while there was no medic in sight. Since it seemed that no medic was there to claim it, my father told one of the men he was with to recover the ruck.

It was only after 49 years that my father got to know the story behind the medic bag when he got to talk to Wilson, the medic. According to Wilson, there was this other medic that went up the captured hill and got knocked out by a large tree limb that fell on him, and that bag that was left on the hill was his. That is another one of those mysteries of the war, the mystery on the unattended medic bag, explained after almost half a century.

On their way to the top of the hill, it was a big surprise to them when they spotted the 51-caliber heavy machine gun. It was a robust weapon, weighing about 1,000 pounds, that were equipped with an armor plate and wheels. The armor plate that provided a shield would explain how its operator had been protected from any of their rifle rounds that may have hit the machine gun.

When my father spoke of the machine gun, he expressed in amazement, "Imagine maneuvering that around in the jungle!" Another surprising thing about the machine gun was the gunner was chained to it. So, the gunner would not be going anywhere when he operated it—they found him dead.

During the battle, one of their men threw a CS tear gas grenade towards the hilltop. My father always thought that was a big help in taking out the machine gun and finally capturing a knoll going up to Hill 724. The knoll they captured was well fortified with bunkers at the hillside. They decided to set up there for their nighttime position.

There were some downsides to the knoll though. The far side of the knoll was not so good. Its fields of fire were poor, and they had no time for making them better. Alpha Company took that side because there had to be a group there. Also, the other side that Delta Company took had gaps in between and did not have a place for a chopper to land because it was a triple canopy jungle.

Their contact with the NVA earlier caused a lot of casualties, which included my father. It was Alpha Company that had the most casualties. The worst off from all the casualties would be a gunshot lieutenant from Alpha Company. He was a replacement that came in for Alpha Company, which made him the second guy that came in in just two days. Then a third one again came to take over for Alpha Company. Even after that third guy in two days, came the fourth guy in just three days, whom my father later met at a reunion after more than four decades.

A medevac chopper came that evening around 10 pm to take the WIAs to a hospital. The gut shot lieutenant was prioritized among the WIAs. The personnel in the helicopter lowered a stretcher for the lieutenant. To bring it down, they used a winch and they also had to weigh it down since it was of flimsy material. As per the Geneva Convention, medevac choppers are not to carry munitions. However, all they had at that time for weighing down the stretcher in order for it to penetrate through the trees were cases of grenades and rifle ammo. It was a joyful sight to behold, my father thought.

Once the stretcher was brought down, the Alpha lieutenant was raised to the chopper and sent to a hospital. The next chopper that came used a jungle penetrator for taking the WIAs. A jungle penetrator can be likened to an anchor. It would be lowered down, then a WIA would have to sit on it, and he would be raised to the chopper and pulled inside. My father was taken by that chopper.

After he went up, the chopper still hovered and hauled up more of the wounded. When he got in there, he was feeling selfish and was hoping it would take off immediately like what the first chopper did. It was out of anxiety that the chopper might get shot down as it hovered. My father would not call it idle fear though. Such a disaster could happen. He even later heard that there were 13 resupply and medical choppers shot down.

He felt relief when the chopper finally took off. Although his mind was finally calm, he seemed to not think straight. He felt like smoking a cigarette. He could not even inhale. Indeed, combat can change one's thinking.

As my father was sent to the hospital, Alpha Company and some volunteers from Delta Company had gone through hell. Since it was a captured knoll that they set up for the night, a counterattack was expected. It was bad enough that Alpha Company had lost a lot of men as they went uphill, what made matters worse for them was that the side of the knoll they occupied got partially overrun by the NVA.

They occupied five of the NVA's bunkers that had been set up with overhead covers in their side of the captured knoll. The NVA attempted to retake those bunkers. A sergeant from Alpha Company crawled over to Delta Company's side and asked for volunteers to come and help Alpha Company in fighting off the NVA that were overrunning their side.

Half a dozen guys, including Byron Kinnan, whom my father had spoken to regarding the incident, just went over. Without being ordered, around six or seven guys willingly started to crawl over to Alpha Company's side. Some others also followed, but others just stayed. They tried to recapture the NVA bunkers that had been overrun by the NVA. When my father's friend, Byron jumped into one of the bunkers, there were five dead Americans and NVA that he stood on. Byron decided to use the dead body of the NVA soldier to reinforce their overhead protection, so he took the body of the dead NVA soldier out and threw it at the top of the bunker.

More NVA also came up the hill. The NVA attacked the bunkers that Alpha Company occupied with machine guns. They noticed that the NVA had tracers in their machine guns.

Typically, it is in every fifth round of a machine gun that is a tracer, which is a bullet with a pyrotechnic composition that would go off burning brightly, would be fired. So, in daylight, the tracer is highly visible, while at night, it comes out very bright. Its visibility helps to let the shooter trace the direction it is going towards, thus also allowing him to know where it is exactly going and make ballistic corrections when necessary. It can also serve as a tool for marking a target for other shooters to fire at. Some of their men's weapons were equipped with tracers too. While there are some advantages to tracers, unfortunately, there is also one disadvantage to it, which is the enemy would also be able to see where the shooter shoots from.

Rifles, too, would emit fire called a muzzle flash. It occurs due to the mixing of the gunpowder's combustion products and remaining unburned powder with ambient air. So, then they would use grenades as an alternative. Using grenades helped keep them unseen. Also, in such situations like the one they were in where the point of origin of the attacks towards them was not exactly seen, it was apt to use grenades.

On the contrary, when the enemy was right on top or just beside your position, it would not be suitable. My father's friend, Byron threw grenades at the NVA's bunkers. Unluckily, the grenades he threw did not get inside the NVA's bunkers. Some just landed on top of the bunkers, which were made of five-inch logs that made good overhead protection. While some just landed and went off outside the bunkers.

Suddenly, one of their men snuck up to Alpha Company's side and he got to throw in either a grenade or satchel charge. Because of that, they were able to recover.

They had no idea, but the battle that they were engaging in at Dak To would become the largest, bloodiest battle of the war at that time. In fact, in just the first four days that my father got involved in the war, 3rd Battalion 8th Infantry already had 46 men KIAs. For every KIA, the projected WIA was four. So, the total casualties could be rounded off to approximately 300 casualties in a 500-man battalion in just four days, in which he was included.

Among the 20 original officers they had, Bell was the only survivor. On just their first night at Hill 724, Alpha Company had a lot of losses when they were ambushed by the NVA when they arrived at the base. Alpha Company suffered even more when they made it up the knoll that they captured in Hill 724. The night of their contact with Charlie at Hill 724, the wounded were gathered in a collection point, where they were then taken by medevac choppers to the hospital.

That was followed by the chopper that brought in a large cargo net full of resupply of ammo. The circumstances at the knoll made the delivery of ammo a struggle that the large cargo net carrying ammo was brought down in the wrong spot.

The pilot could not be blamed though because when he came to send them a resupply of ammo, they were being counter-attacked. Also, the chopper brought the resupply amidst a triple canopy jungle with no landing zone in the darkness of the night. The following morning, the cargo net that brought down the ammo was used to haul out the dead.

Within the first three and a half days of the battle, the remaining troops of the Alpha Company had been evacuated and Bravo and Charlie Companies had been appointed to replace them.

While the Battle of Dak To raged on, my father was being treated for the serious wounds he had been inflicted within the battle. He was hospitalized in-country for three weeks. He was initially accommodated in the 71st Evacuation Hospital. He was later transferred to the Cam Rahn Bay Hospital. Then he was sent to a hospital in Japan, the Japan Army Hospital in the Tachikwa AFB, Kisini Barracks, where he spent a month for recuperation.

There, he met a helicopter pilot who got evacuated

because he had not been able to crap for a month. He was literally scared shitless. Every day, the medics came to him with an ice teaspoon to aid in moving his bowels. My father did not know how it ended, but he made two comments on the situation: 1. Do not buy surplus army ice teaspoons. 2. There is truth to the phrase "scared shitless."

It was on December 30, 1967, when my father was finally sent back to Vietnam. He was given a new pair of fatigues. In the duration he spent at war, he had lost some weight. He came at 175 pounds with a 36-inch waist. By December, he became 135 pounds with a 28-inch waist. He became 40 pounds lighter even after he got fattened up when he stayed in the hospital for seven weeks. When my father got back to Vietnam, since he was wounded, he was offered the job of escorting celebrities of the likes of Swedish-American actress, singer, and dancer, Ann-Margret Olsson, who is well-known magnanimously as Ann-Margret.

He did not want to take it easy and wanted to get back to real work. He remarked when he recounted the job offer, "Like a fool, I insisted on returning to my 6/29 Arty-3/8 Inf unit."

Early in the battle at Hill 724, my father received two Purple Hearts. It was only 50 years later that he received a Silver Star. It did not get to him at the time of the war, as he had been sent to Japan for recovery.

When you are serving in the war, there is a high chance you are going to get wounded. A friend of my father, Byron Kinnan, who also fought in the battle of Dak To, shared to my father his own experience when he, too, had been wounded in the battle in Hill 724 that also affected his lungs. He had collapsed lungs, which caused his lungs to be filled with blood.

When he had been laid on a gurney, he was greeted by a doctor who introduced himself as Lieutenant Colonel Major. He thought, "What kind of new rank is that?" He found it very confusing as the doctor had the rank of lieutenant colonel and his last name was Major.

What the doctor did was let Byron sit up to be treated for his collapsed lung, which according to Byron was a painful process. The doctor cut about three-fourths of an inch on his back with a scalpel. In the slit he made on Byron's back; he pushed a plastic hose in about a foot deep. Then it was the medics who resumed the pushing of the hose. They pushed the hose in to penetrate the chest wall membrane to collect the blood.

So, they did that until the hose broke through. Attached to the hose for collecting blood from the lungs was an empty cider bottle since back in 1967, plastic bottles were not yet commonplace, and the hospital had none. They used adhesions for the hose, and he felt as though his breath was taken away when the medics yanked the hose out of him about two to three weeks later due to the adhesions.

CHAPTER EIGHT

From the Firebase to the Frontline: A Soldier's Journey Through Adversity

When my father transferred from Delta Company, 3rd Battalion, 8th Infantry to Bravo Company, 3rd Battalion, 12th Infantry later in their mission in Dak To, they were building a new firebase (FB), the FB25. The FB25 was the 6th Battalion, 29th Artillery FB, which was under the command of a southern gentleman, Capt. Robert Barret, Bravo Company, 3rd Battalion, 12th Infantry.

My father found building a firebase a great duty because they did not have to deal with mountain rucksack humping. They only had to bear with the H&Is being fired at night that could interrupt their sleep.

Every other day, either of the two companies in the FB would go out to do a quadrant sweep, which is going about a quadrant of the compass around two to four clicks (km) away from the FB to see if there were signs of Charlie sneaking around to attack them. They would go out in the morning without a rucksack, and when they came back at night, they would probably cut down or blow up trees to create an alternate landing zone (LZ).

January 19, 1968 was one of the days in the war my father remembered very well. It was Bravo Company, 3rd Battalion, 12th Infantry's turn to do a quadrant sweep. By then, it had already been almost three weeks that my father had been back in Vietnam from recuperating in Japan. The night before, my father and his recon sergeant, SP4 Timothy Nafe, gave each other haircuts with a pair of manual haircutters Nafe had. My father recounted from his experiences, how even something as simple as a quadrant sweep can be dangerous. Just a few days earlier, they had done a quadrant sweep, and my father got into an accident.

While they were doing a quadrant sweep, they were also cutting down trees to make a clearing for an alternative LZ. As they were cutting down trees, my father stopped and watched a tree falling, and he noticed that some of the troops he was with were motioning at him. He was baffled why because the tree was not going to fall in his direction. Then, "Thud!" A large tree limb right above him that was dislodged fell on his head.

"I saw stars during the daylight," he remarked. It was a good thing he was wearing a helmet, or his head could have been split open. Thankfully, the army does concern and care for its men. They sent my father to an aid station via a helicopter ride for free to get an x-ray done on his head. The x-ray result showed no problem, so he went back to the firebase.

Concerning gestures, there was also this one other time my father recalled, which was in the last phase of Ranger School in March 1967 in Florida, when some troops gestured at him wildly, but he could not discern what they were trying to tell him.

They had come across this wide but not too deep stream that was very cold since it was winter at that time. He was ordered to go across it and reconnoiter the area on the other side. By the time he reached the other side, they gestured at him to come back with a sweeping movement. Then he went straight back to the side where he came from.

On his way back, they advised him to only pass by the knee-deep stream. When he returned, they told him that the reason they were gesturing at him was that they saw this large cottonmouth snake that was basking on a rock near him. He was that close to almost touching it as he was coming and going around the area. So when out in the field, it is important to pay good attention to gestures.

Anyway, the reason why January 19, 1968, was engraved in my father's mind was that it was another chaotic day, wherein they once again had an encounter with the NVA. So that day started with that little walk in the jungle as usual. To their relief, without the burden of having to carry a heavy rucksack that time. They just had with them their web gear, harness, canteens, ammo pouches, rifle, extra 20 mags of ammo, and some C-rations for a picnic lunch.

After walking about four clicks (kilometers), they had reached Hill 800, and the soldiers walking point (the leading soldier advancing through hostile territory) spotted a couple of North Vietnamese Army (NVA) trail watchers. They sat at the cross point of two trails. Those trails were their backyard, and they were at home.

The trails were critical to the NVA units, with trail watchers directing troops bringing in food, ammunition, and medical supplies. When the infantrymen at the front spotted these trail watchers, they radioed their commander for guidance. Both Captain Robert Morton and McNulty responded, "Shoot them." McNulty cautioned the captain, noting that the situation could deteriorate rapidly, and decided to call in artillery support.

While waiting for the artillery, the infantrymen took the opportunity to rest and eat. McNulty used this time to train Specialist Fourth Class Timothy Nafe on how to call in artillery fire. Nafe, around 20 or 21 years old, had previously expressed confusion about why they were fighting in Vietnam, and McNulty had tried to explain the Domino Theory to him.

Since McNulty hadn't adjusted artillery fire since returning from an injury, he felt it was urgent to cross-train Nafe. He taught him how to locate their position on the map and translate the coordinates into code—a precaution in case the enemy captured their maps. They began by calling in a smoke round 500 meters from their position to confirm accuracy.

Hearing the distinctive "pop," they proceeded to adjust the fire closer in 100-meter increments, then 50-meter increments when within 300 meters, which is considered "danger close."

They also ensured all six artillery tubes at the firebase were aligned correctly. By coordinating the timing and adjusting the aim, they got all tubes to hit the same point. Satisfied with their setup, McNulty informed Captain Morton, "We're good. We've got artillery."

Unbeknownst to them, there was an extensive NVA basecamp 70 to 90 feet below their position, complete with tunnels, field hospitals, and facilities to support hundreds of soldiers. Although Bravo Company knew there were NVA forces in the area, they were unaware of the base camp's size and the concentration of enemy soldiers.

Bravo Company, 3rd Battalion, 12th Infantry, found themselves deep in the NVA's territory. The enemy was well-prepared, uncovering foxholes, setting up fighting positions, and placing snipers high in the trees. The NVA were adept at operating in the dense triple canopy jungle—a layered forest environment that limited visibility from the ground to just 15 feet above. However, snipers in the treetops had a clear view through the lower foliage.

McNulty had no idea that the NVA routinely operated from the trees. His experience hadn't revealed this threat, and he hadn't been in the area long enough to learn it from other soldiers in Bravo Company. After a year of service, soldiers were rotated out, and replacements spent only a week or two learning from veterans. It was only decades later that McNulty realized this was part of the NVA's tactics.

Captain Morton ordered his men to move out. As Tim Nafe stood up and took his first step, he was immediately shot through the heart and died before hitting the ground. He was likely targeted first because he was using the radio and directing artillery, making him an obvious priority for the enemy.

In battle, anyone identified as essential—like those carrying radios, pistols, carbines, or wearing officer insignia—is at high risk. Eliminating leadership and communication personnel can render a unit directionless and easy to defeat.

Bullets began flying everywhere, like a swarm of hornets kicking up the ground around them. The radio was still with Tim's body, and McNulty urgently needed it. He shouted to someone nearby, "Throw me the radio!" The captain tried to report their situation to higher command, but they didn't grasp the severity. Frustrated, Captain Morton exclaimed, "I'll tell you how serious it is. I'll put a dead man on the phone to tell you!"—referring to Nafe.

Amidst the chaos and mounting casualties, McNulty needed to act quickly. He adjusted Nafe's artillery concentration closer onto the NVA attackers to block reinforcements. However, with bullets coming from all directions, identifying enemy positions was a struggle—it felt like they were in the middle of a circular firing squad.

As McNulty adjusted fire from the firebase, he received a call from another unit offering assistance with 155mm howitzers. He directed them to fire further out to block more NVA reinforcements. Then, a 175mm unit also offered help. Despite jokingly replying, "No, I'm kind of busy now," he was grateful for their support. By then, he was coordinating fire from three different artillery batteries.

He faced a situation not covered in Artillery School: utilizing fire from multiple batteries while on the move and trying to escape an ambush. Typically, coordinating several batteries is done by higher-ranking officers and involves firing at a fixed target—not a moving one.

Unable to pause and make detailed calculations, McNulty adapted on the fly. Another unit with 8-inch howitzers joined in, bringing the total to four batteries under his direction. He decided to use the larger artillery for covering and blocking fire to protect their withdrawal and prevent NVA reinforcements.

He provided each unit with coordinates forming an inverted "L" shape. The long side protected their withdrawal route back to the firebase, and the short side prevented the NVA from flanking them. As they moved, he adjusted the "L" closer to Bravo Company, allowing more movement back to safety.

After-action reports suggested only a couple of reinforced NVA companies were attacking, but from McNulty's perspective, it felt like the entire North Vietnamese Army. The NVA tried to get within 25 meters of the Americans to avoid artillery fire—a tactic known as "hugging" the enemy. Beyond that range, they were vulnerable, assuming the forward observer was still alive.

Over four grueling hours, Bravo Company withdrew to the firebase, enduring ambush after ambush. McNulty found himself alone, not realizing how it happened as he was preoccupied with adjusting artillery fire. He had positioned himself at the rear to avoid bringing fire too close to his unit.

At one point, two GIs without rifles approached him, asking for his weapon. They had dropped theirs while carrying a wounded soldier who died. Since McNulty hadn't used his rifle, focusing on the radio and map, he agreed to lend it to them in exchange for a grenade, on the condition they stayed with him for protection. They agreed, but within minutes, they were separated in the chaos.

Although alone, McNulty received reliable support from the artillery units. As they neared the firebase, he switched from 105mm guns to mortars, better suited for shorter ranges. He noticed mortars falling in their area and feared they might be friendly fire, recalling a previous incident with short rounds. He ordered a check fire on the American mortars, but the shells continued—it was enemy fire. Relieved that it wasn't friendly fire, he resumed directing the mortars.

The horrors of combat weighed heavily on McNulty. Nearing the firebase, he felt desperate enough to consider injuring himself to escape the frontline for a few days, but he pressed on. Within 100 meters of the firebase, he encountered a major in a clean uniform who offered him a drink from his canteen—ice-cold scotch, which McNulty didn't like.

Exhausted, he yearned to reach the firebase and collapse.

Upon returning, he still needed to hand over artillery coordination to the major. When someone asked if he had done it already, he went back out to the trail to complete the handover.

The battle at Hill 800 left a harsh aftermath for their unit with several casualties and a lot of equipment lost. As McNulty heard initially at the firebase, the casualties from the battle due to the fog of war were around 17 killed in action (KIA) and 35 to 40 wounded in action (WIA) out of around 75 men. After the battle, it was only Nafe's body that had been brought back. The battle they fought was intense.

Niceties like bodies were of low priority when the casualties are high. Only two radios—his and the captain's—got back to the firebase. Not even one machine gun got back. A lot of weapons, too, were lost; they may have either been jammed or the soldiers may have dropped them to carry the wounded.

Fortunately, it was later discovered that the death toll from the battle for their unit was only 7, not 17. After the battle, every day, the infantry would search in and around

Hill 800 to recover and bring back the other six KIAs. Also, their B52s carpet-bombed Charlie's area after the battle. That caused extremely high losses for Charlie.

However, those that were inside the spider holes or made it inside in time before the bombs were dropped, to their luck, were spared from the bombing. The spider holes 90 feet underground were left untouched. So those that survived the carpet-bombing soon left the area. Their infantry also resumed searching for their dead men, even after they had carpet-bombed the area.

Although they managed to subdue the NVA, naturally, they knew that a counterattack would follow. At any moment, they expected it to come. After my father finally returned from the field and turned over artillery to the major, he went to the command post bunker where he left his rucksack. He found it nice that by his rucksack, his rifle and ammo that he lent to other soldiers had been returned.

On that same day of their encounter with the NVA, mail caught up to my father. He got the news that his friend was killed. That friend of my father had been thrown out of the car and got his head hit on a fire hydrant. My father found it ho-hum. As my father explained, "Everyone has a bad day. I was just numb from the day." He also found out that his brother was on his way to serve in Vietnam. He was not so thrilled about it. He thought, "If only one can be in the theater at a time, I'm going."

An hour or so after getting mail, my father was called

into the command post and got good news. He was informed that his Air Observer School application was approved. Air observers (AO) are those who would fly around in a little two-man helicopter. In the helicopter, there would be a pilot and an artilleryman that would be directing fire as they are flying. My father's friend, Larry "Shot-In-the-Back-of-the-Neck" Skoglund, who had replaced him after being wounded twice at Hill 724, became an air observer, himself. While he was doing his job as an AO, he also cross-trained on flying a chopper.

Learning how to fly a chopper eventually became helpful for him because one time, when he was aboard a chopper, the pilot flying it got shot and died. Larry was able to put his knowledge and ability in flying to good use. However, Larry also got shot. He was doing fine until he became unconscious due to blood loss and crashed.

Fortunately, he was able to survive the crash. He wound up at the 71st Evacuation Hospital, where my father had also been sent.

Although it was strange that they said at the command post that my father's application for Air Observer School was approved. Because he never applied as an air observer (AO). He just did not bother arguing with them when they said that. He believed that it was just their way of being nice and polite to him and rewarding him in recognition of his work as a forward observer (FO). That made him think of the other advantages that would come along with attending AO School—like he would be able to sleep in a cot, maybe even have sheets too, eat real mess hall food, and maybe get to shower weekly—it sounded like heaven for him. It was a way to finally escape from the field.

Someone had made other plans for him, which would provide him a faster way to finally enjoy a bed and sheets. After my father had been informed about AO School, someone opened up one of their 50-caliber machines, as some NVA have been spotted near their perimeter. The incident brought back that feeling of blood instantly pooling in the stomach. It was natural for a counterattack after an encounter with Charlie. So, they anticipated it that night.

At that time, they still had not finished the construction of their firebase. The area where the firebase was situated had been a forested hilltop. Since there so many trees around, there was a lot for them to cut down and to clear out for fields of fire.

As my father recalled, they had a Caterpillar tractor and a backhoe. It took a herculean effort in removing trees, digging out holes for howitzers, sandbagging, and parapetting. Since the firebase was not finished yet, they had a big shortage of bunkers with overhead cover. So, the bunkers would be crowded.

In the bunker where my father was, probably over 20 men occupied it. It was so cramped, that the occupants would sit in this folded-up position that had their knees up to their chest, and there would be a guy sitting behind them. The bunker was rather cramped, and my father felt claustrophobic.

Their situation in the bunker reminded him of an activity they did in Ranger School. It was a wake-you-up exercise. In the exercise, there were these approximately twenty-something feet truck tires that were supported by a steel cable tied between two large trees that they had to crawl into with their rifle in hand. They would each enter after the guy in front of them did. It may sound so simple but not quite.

There was only a little difference from the width of the tires to their shoulders, which gave only a couple of inches of space for them to move through. Their vision would be blocked by the guy in front of them. Although they still had light coming from behind them, the light would shortly be blocked as well when the next guy to enter followed.

My father related that the exercise can bring one's heart into overdrive. As they went in the tires, there was no light in front of and behind them, their shoulders touched the inside of the tires, their rifle would catch on the tires, and the trap-like obstacle of tires they crawled through swayed. All of that made it quite a challenge passing through, and they knew that they had better not make any noise.

It was only when the guy in front of them had exited the obstacle that they could finally see light. Finally, by then one's heartbeat would return to normal.

My father thought if a counter-attack would soon come that evening, he would not be able to do a thing if he was trapped in one-third of the way in the crowded bunker, so he decided to get out of it. It was "Good-bye tire swing." Among the men occupying the bunker was an RTO positioned in the front. The RTO was in contact with the listening post (LP) men. My father informed the RTO that he was going to move to the outside edge of the bunker and requested for him to be called when there was any report from the LP. The major counterattack they anticipated did not come that night, but a smaller one, seemingly directed at my father did. Three 82 mm mortar rounds had been dropped in their perimeter by the NVA just to wake them up.

Unfortunately, one of the mortar rounds exploded at an arm's length from his head. The impact of an 82 mm mortar is equivalent to that of seven hand grenades. If it was not for my father's sixth sense, the injury he suffered would have been more serious. His sixth sense had warned him in about one-fourth of a nanosecond right before the mortar exploded. He had his forearm placed across his forehead. The explosion would have hit and taken off the top of his head.

Instead, the mortar injured him in other parts of his body. The mortar hit his arm and took off a baseball size chunk out of it. His arm also got broken in a few places. Two of the three nerves in his hand and fingers were also cut. His eyes, too, got inflicted. Both his eyes were open. He lost an eye and almost lost the other eye from the shrapnel that was about one-sixteenth away from his eyeball. His eyelids were left unharmed since his eyes were open. The mortar also made a hole in his leg by the femoral artery. It also blew out one of his eardrums.

Just one other problem for him was the shrapnel that hit him. There were about 40 pieces of 1 to 2-inch shrapnel and 400 smaller pieces. It was a miracle that he survived the explosion that was just an arm's length from him. "My nickname is Lucky," my father even said. Indeed, that nickname was apt for him.

My father was not even unconscious. He, somehow, was able to get up, but he had a problem in seeing. He was not aware at that time that it was blood in his eyes that hampered his sight. He also tripped and was not aware of the problem with his arm. When he got up, he went to the front of the bunker and asked the men inside if there was a medic around.

Someone answered, "Yes." Then he said, "Come outside. I've got some work for you." That was another one of those instances that showed that combat can change your perspective.

Due to the shrapnel, my father was flown to an aid station. The medics had taken several big pieces of shrapnel out of him. Then he was sent again to the 71st Evacuation Hospital to get further treatment. That time around, he got to have the drop on Lt. Col. Major. He greeted him with a "Hello," and asked if Major still remembered him. He was then taken care of in the operating room.

The following day, Major informed him that they needed to en-ulcerate his eye. His eye was just like jello. Major also told my father that the war was over for him. He was overjoyed with the news. He was in such a bad condition that he could no longer resume fighting in the war and was to be sent home.

His eye was almost swollen shut, but he kept closing it and lost count of how many times he closed it out of curiosity on what seeing with only one eye was like. It may have seemed kind of foolish, but my father explained that life and reality are variable. Around that time, my father received his third Purple Heart. My father spent a while in the 71st Evacuation Hospital for recuperation. He stayed there until he regained enough stability to move.

Ten days after he was admitted to the hospital, word filtered down about how the South Vietnamese guards at the hospital stopped and arrested one South Vietnamese helper whom they caught around the area pacing off distance from the hospital's flagpole to another building. The help had been paid by the enemy to count her steps for them to know the distance required in firing their bombs for attacking the hospital.

The South Vietnamese workers at the hospital had been alerted that the enemy had plotted an attack on the hospital, so in the afternoon, no workers reported for duty. The NVA also mortared and rocketed the hospital.

They also breached and partially over-ran the hospital by a ground attack. Around that time, my father also received his fourth Purple Heart.

His friend, Larry "Shot-in-the-Back-of-the-Neck" Skoglund was also there in the hospital. He also got two Purple Hearts for when he got shot in the back of the neck and when he got shot in a chopper he was aboard that also crashed.

After spending a few months recovering at the 71st Evacuation Hospital, my father was stable enough to travel to the United States for further treatment at Walter Reed Army Hospital. He stayed at Walter Reed for a year and a half, during which he made several close friends.

One of these friends was Max Cleland, who later became a U.S. Senator from Georgia. Cleland was at Walter Reed after a grenade accident while serving in South Vietnam. As he was getting off a helicopter, a loose pin fell out of one of his grenades. The grenade hit the ground and exploded, resulting in the loss of two legs and an arm.

My father was very careful with his grenades, often wrapping electrical tape around them to keep the pins secure. Cleland wrote a book about his Vietnam War experiences called *Stronger at the Broken Places*, where he mentioned my father. In the book, Cleland nicknamed him "Weird Harold" because of the creative stories he told about how he was injured.

Another friend my father made at Walter Reed was Bob Nadolski.

Nadolski lost an eye and had a leg amputated above the knee. They became friends when an army nurse, Major Barker, introduced them.

She suggested, "Hey, we just got a new guy, Nadolski. Why don't you go talk to him?" My father approached Nadolski, and they found they had a lot in common. They soon started doing activities together, like fishing.

During their recovery, both men underwent surgeries and spent 10 to 15 days in the hospital ward afterward. After each stay, they went home on medical leave where they could relax and do what they wanted. Then, they returned to the hospital for more operations. This routine gave my father and Nadolski time to spend together during surgeries. They often went fishing or participated in other activities, deepening their friendship over the years they shared the same ward.

Eventually, Nadolski was referred to the Physical Evaluation Board (PEB), which assesses whether a service member can continue to serve based on their injuries. Nadolski was rated with a 90% disability and was discharged from the Army. My father was also referred to the PEB and received a 93% disability rating. Although the board recommended he retire, he chose to remain on active duty. He returned to Fort Sill, where, after six months, he became a battery commander and battalion adjutant.

Reflecting on his time in Japan, my father found the Japanese people to be kind and friendly, without any lingering animosity from past conflicts. He admired how clean and well-maintained their streets and harbors were. He remembered getting a traditional straight-razor shave at a Japanese barbershop while still recovering from his injuries. As he sat in the barber chair with his face covered in lather, he realized that the U.S. and Japan had been enemies just one generation earlier during World War II.

He wondered if the Japanese barber, who was about to shave him, had lost any relatives during the war and if this might be "payback time" for her. Fortunately, his fears were unfounded.

He also recalled meeting a Japanese hospital worker from the Japan Army Hospital. When she saw him again, she looked worried and asked, "Haven't you just been here a couple of months before?" She was upset that people were experiencing such hardships once more.

My father was deeply moved by how the Japanese treated him and others. Despite being enemies a generation earlier, they showed kindness and respect. He believed that the U.S. demonstrated its greatness by treating former adversaries with such goodwill.

He experienced similar warmth and friendship during his visits to Germany. There was no hatred towards the U.S., only respect. He also noticed that Germans loved American music from the 1960s and 1970s, which made up about 30% of the music played in the country.

My father's friend Nadolski also felt welcomed when he took several vacations to Vietnam, visiting families he had met before.

CHAPTER NINE

A Soldier's Return: Resilience, Redemption & a New Mission

After the Vietnam War, my father faced new challenges that tested his resilience in different ways. Despite his disabilities—a 93% disability rating from the Physical Evaluation Board (PEB)—he chose to stay on active duty instead of retiring medically. In June 1969, he returned to Fort Sill, where he served as a service battery commander. Six months later, he moved to artillery battery commander.

Away from the horrors of war, my father encountered a different kind of struggle. The battery he commanded was often assigned the most troubled soldiers, many of whom frequently went absent without leave (AWOL). The First Sergeant would sometimes ask, "Do I pick them up at the jail or the penitentiary?"

Realizing that a new approach was needed, my father decided to treat his men more fairly and with respect. His fair treatment made a significant impact. The men in his battery stopped going AWOL and achieved a record of three consecutive 60-day periods without any AWOL incidents. My father celebrated this success by saying, "We've had 60 days without an AWOL. Why don't we have a Saturday off?" The soldiers appreciated these well-earned breaks.

On a few occasions, he organized field day picnics, bought hamburgers and hotdogs and had chefs cook for everyone. The unit maintained an AWOL-free record for six months. When new soldiers joined, the existing men would warn them, "Don't even think of ruining what we have. We'll make sure you don't mess this up."

After about a year as a battery commander, my father's role shifted to transferring ammunition to the field and training men in artillery firing. They spent almost a year living in the field, which my father found manageable. They were out in the field six out of seven days or thirteen out of fourteen days. When they returned for weekends, my father insisted that his men set up their tents immediately—using standard military tents. This was practical because the tents would be wet and heavy from rain, and setting them up right away allowed them to dry. Other batteries would leave their equipment stored, only to set up their tents later and get muddy again. My father's men, already prepared, would watch and laugh as others struggled.

An announcement came that representatives from the Department of the Army (DA) would be visiting to discuss future career plans. My father wanted to attend the Career Course, the next step in training for someone with two and a half years as a captain. However, during his discussion with the DA, they noted, "We see you haven't been to Vietnam in four years or so." He replied, "Yes, I was wounded four times and given up for dead three times. I don't think I want to go back there."

Instead of granting his request, they offered, "Maybe we can arrange a 13-month unaccompanied tour to Korea." Not interested in this idea, my father thought, "I've served my time and paid my dues. I'm not looking forward to that at all. I'll take my 90% disability and just retire." This decision led to his medical retirement.

Many American soldiers believed they had won the war on the battlefield, but the overall outcome was different, as North Vietnam ultimately prevailed. As the U.S. began withdrawing its troops, a small number of American forces remained as a symbolic presence for a few more years.

After the war, Congress decided to cut off military aid to South Vietnam, including not stopping supplies sent through the Ho Chi Minh Trail—a network of routes through Laos and Cambodia named after North Vietnam's leader.

About a year and a half after American troops left, North Vietnam launched a rapid offensive, pushing southward so quickly that many South Vietnamese citizens fled in panic. With only a limited number of helicopters available, the remaining Americans could evacuate only their personnel.

North Vietnamese forces were determined and well-led, allowing them to break through South Vietnam's defenses. Compounding South Vietnam's problems was widespread corruption among military leaders, local chiefs, and police. Many were found stealing weapons meant for the troops and selling them to the North Vietnamese or Viet Cong, leaving South Vietnam's forces ill-equipped and accelerating their defeat.

Meanwhile, many American soldiers suffered greatly during the war, with many left injured or disabled. Returning to the U.S. did not bring them peace. Instead, Vietnam veterans faced a harsh reception at home, often met with hostility—spat on, insulted, and even physically attacked.

The emotional and psychological toll of the war, combined with this treatment, led to significant trauma for many veterans, marking the beginning of a different kind of struggle.

Unlike the honor and high regard afforded to World War II veterans, Vietnam War veterans were subjected to disdain. They had sacrificed their lives and endured the terrors of war, yet were treated poorly upon returning home. Instead of being honored, they were seen as symbols of national failure and became objects of disgust and anger.

The Vietnam War stretched on for nearly two decades, fueling growing protests and widespread doubt about America's involvement. Unlike the triumphant homecomings of World War II, returning Vietnam veterans—many barely out of their teens and drafted into a conflict they never chose—were swept from the jungle's chaos to U.S. soil in a matter of hours, with no time to process what they'd endured. There were no welcome-home parades, no weeks aboard ships to decompress, and no collective demobilization; instead, soldiers stepped off the plane into indifference, hostility, or outright scorn. Many were even warned not to wear their uniforms in public, for fear of harassment.

In Vietnam, the constant threat of death forced soldiers to shut down emotionally—tomorrow, a brother in arms might be gone. That survival mechanism went unaddressed back home, where trauma was poorly understood and therapy was stigmatized. With scant resources and few willing listeners, veterans buried their pain: the boredom punctuated by terror, the sudden loss of friends, and the split-second decisions that haunted them. But buried trauma doesn't stay hidden; it erupted in nightmares, flashbacks, anxiety, depression, and substance abuse, as many turned to alcohol to numb memories that replayed night after night.

Despite being shamed by much of society, most veterans remained quietly proud of their service to the U.S. and South Vietnam—and deeply resentful of those who attacked them for it. Their legacy is not confined to history books but lives on in their bodies, minds, and hearts. Recognition, healing, and understanding—which were so tragically delayed—still matter today. It's never too late to honor their sacrifice, not just with medals, but with truth, dignity, and compassion.

Another unpleasant experience for many Vietnam War veterans was society's lack of preparedness in providing them support for readjusting to civilian life. After World War II, the G.I. Bill was established, offering generous financial assistance to veterans for living and educational expenses. However, Vietnam War veterans did not receive the same generosity. The enormous cost of the war strained government finances, and Vietnam veterans received only $200 per month, barely enough to support their needs and education.

America's economy was also struggling, facing an economic crisis and stagflation. Fortunately, my father did not face mistreatment when he returned home. Frequently, veterans faced hostility at airports or when in military uniform in public. My father returned via military plane and bus, so he did not encounter these problems.

However, his friend Bill Ferguson was one of the unlucky veterans who were mistreated upon returning. Angered by those who were rude to him, he would physically confront them, sometimes getting into fights and trouble with the police. His mother advised him that since he had dual citizenship—including Australian citizenship—it might be best to leave the country before he got into more serious trouble. Following her advice, Bill moved back to Australia to escape the hostility he faced in America.

After the war, my father developed new routines that included spending time with family and friends and expressing gratitude to those who had helped him. He would often take the family for walks in Washington, D.C. During that time, there were race riots, and areas in cities like Baltimore and Detroit were burning.

On one of his trips during a convalescent leave between operations, his car was broken into in Baltimore, and his radio was stolen. After that incident, for security, my father obtained a license to carry and started keeping a .38 revolver in his car.

One of the things my father did was send a thank-you check to Tim Wilson, the medic who treated him when his lungs were punctured by grenade fragments on Hill 724. He sent the check every November 7—the anniversary of his injury—so Tim could enjoy dinner out with his wife. Although my father never confirmed it was Wilson who treated him, and Wilson couldn't recall for certain, my father felt it was important to show his gratitude.

Another friend my father stayed in touch with was Capt. Terry Bell. They kept in contact by phone and would visit each other, when possible, often meeting at battalion reunions.

At some point, my father felt compelled to reach out to Specialist 4 Tim Nafe's mother, about ten years after Nafe had been killed in the Battle of Dak To. He visited "The Wall" in Washington, D.C.—the Vietnam Veterans Memorial—and learned where Nafe's hometown was. It was just an hour and a half away. He and my mother visited Nafe's mother and shared with her how her son had died. Finally, knowing the details lifted a burden she had been carrying; at her son's funeral, the casket was closed, and she had feared he might have been captured and tortured.

My father couldn't understand why the funeral was closed-casket since Nafe had been killed by a single bullet to the heart, and his body had been brought home promptly.

For several years, around the anniversary of Tim's death on January 19, my father and our family would visit his mother, and we would go out to dinner together. Eventually, my father realized that if he had been the one on the radio the day Tim was killed, it might have been him who was shot. He felt that if he had been in Tim's place, it's possible that instead of 28 wounded in action (WIA) and seven killed in action (KIA), all 75 men of Bravo Company, 3rd Battalion, 12th Infantry might have been killed that day. This thought gave my father some peace.

After medically retiring from the U.S. Armed Forces, my father looked for a job and tried to buy a house at the same time. Despite his skills and experience, he wasn't getting hired because of his disabilities—being deaf in one ear and blind in one eye made him a risky candidate in manufacturing environments where accidents could happen. He decided to pursue further education by studying for a Master of Business Administration (MBA). Using the G.I. Bill, he enrolled at Temple University in 1974 and completed his master's in less than two years.

He was hired immediately by Ford Motor Company, where he worked in industrial engineering, even though he wasn't formally trained in that field. While working full-time, he single-handedly designed and built an addition to our house, essentially doubling its size. He did everything himself except for the excavation and pouring of basement walls. He planned and built every detail, including wiring, plumbing, installing drywall, windows, hardwood flooring, and roofing.

Eventually, he was promoted to Senior Industrial Engineer after years with the company.

CHAPTER TEN

Lessons from a Life of Duty: Resilience, Service & Legacy

My father has always felt a strong sense of duty to his country. He often said, "As we live in a country as good as ours, it's only right to donate two years of service." He believed that joining the armed forces was about serving, learning, and broadening one's horizons. "People can learn a lot from going into the service," he'd tell me. "You get a chance to see the world and understand how others live."

During his time overseas, he witnessed firsthand the stark realities of poverty. He spoke of places like Korea, the Philippines, and Vietnam, where people lived in shacks and struggled daily. "You see how rotten, miserable, poor, and disadvantaged the vast majority of the world is," he would reflect. These experiences reinforced his appreciation for the opportunities available in our country.

However, when it came to war, his perspective was clear and firm. "No one should ever have to go off to war," he insisted. "War should be avoided and used only as a last resort." He saw war as a failure of diplomacy. Reflecting on Vietnam, he pointed out, "Look at how it ended—it wasn't a military victory. It was people sitting around a table deciding what to do. They could have done that with economic sanctions years before."

His disdain for war was deeply rooted in his own experiences. "Being in combat is the most horrible feeling in the world," he confided. "Wars should only be fought by the politicians who start them."

As a forward observer (FO) in Vietnam, my father held a role that was both critical and dangerous. His responsibilities included directing artillery fire—a task that required precision and nerves of steel. "An FO has more firepower than an entire unit," he explained. "I could be firing six guns, dropping around 40 to 200 pounds of explosives at will." One harrowing night, his unit was ambushed while on patrol. With enemy forces closing in, he had to call in artillery strikes dangerously close to their position. "I brought explosions within 25 yards of us," he recalled. "Anything outside of that was destroyed. It was the only way to keep us alive."

From his military service, he gleaned several life lessons. One of his maxims was, "Anything in excess is a good starting point." This meant there was no such thing as too much ammunition in the field. "We couldn't get a resupply instantly," he said. "So, we had to make sure we had more than enough."

Returning home, my father faced the silent battles that many veterans endured. He realized the average person couldn't grasp stories he kept to himself for decades. Only with fellow veterans did he share his experiences.

When asked how long it had been since he left Vietnam, he often replied, "Just yesterday." The memories were vivid and haunting.

His struggles weren't just emotional but physical as well. He suffered from terrible migraines, a consequence of post-concussive syndrome from mortar explosions. I remember times when he had to leave work at the Ford Motor Company because the pain was too intense. One incident remains etched in my memory: Walking with my mother near Walter Reed Army Hospital, we heard a car backfiring loudly. Instinctively, my father threw her to the ground and shielded her, thinking they were under attack. It was only after a few moments that he realized they were safe.

Despite these challenges, my father emerged with a profound appreciation for life. "I'm just grateful to be a survivor," he often said. "Many soldiers didn't make it back from Vietnam. Life becomes a little more precious when no one is trying to take it." This newfound gratitude fueled his determination not to waste a moment. "I didn't want to sit around wasting my life," he told me. "I decided to make something of myself."

One of his guiding philosophies is, "If you don't know what you're doing, don't do it." He believed tackling a task without proper knowledge was a recipe for failure. "If you don't know how to build a house, don't do it," he advised. "There's a lot that can go wrong." This mindset led him to become a lifelong learner. Before starting any project, he would immerse himself in learning everything he could about it.

His thirst for knowledge was inextinguishable. He became an avid reader of repair manuals and constantly sought ways to fix things on his own. He did everything from repairing vehicles to building a large addition to our house—complete with carpentry, plumbing, electrical wiring, roofing, septic systems, solar panels, and geothermal installations.

He even taught himself how to trim a horse's hooves and shoe a horse. "Nowadays, it's just as easy to learn by watching YouTube videos," he grinned. "You have so many resources at your fingertips."

During his time in Ranger School, he instilled in himself the belief that one should never take the easy path. "Push yourself beyond your comfort zone," he would urge. "Most people take the easy way out and stop working when it seems like a good time. But when you have a job to get done, it doesn't mean stopping when it's convenient. It means staying until the job is done."

I recall countless nights when he'd be outside in the freezing cold, changing the brakes on his car, sometimes working past midnight. He applied this relentless work ethic to everything he did, always striving to work longer and harder than anyone else.

Another principle he lived by was evaluating what should and shouldn't be done. "Don't do what you don't have to," he often said. "There are things that aren't important and don't need to be done at all." However, his philosophy was clear for the tasks that mattered: "Do what you can when you can."

He admired the motto of the United States Navy Construction Battalion—the Seabees—which is, "The difficult we do now. The impossible takes a little longer." He saw it as a testament to the power of attitude and perseverance.

"Whether you say you can or you can't, you're right," he'd tell me. "Impossible is just a word that can be overcome anytime."

Two other philosophies guided his actions. First, he believed in leaving things a little better than he found them. Second, he constantly asked himself, "Have I done all that I can?" He wouldn't rest until he could honestly say, "I did the best that I could."

His generosity was evident in his charitable contributions and his commitment to improving the world around him. He truly wanted to leave the world better than he found it.

Growing up with my father, I was deeply influenced by his values and philosophies. His resilience in the face of adversity, relentless pursuit of knowledge, and unwavering work ethic shaped my outlook on life. He taught me that challenges are opportunities for growth and that the so-called "impossible" can be achieved with determination and effort.

In reflecting on his life, I'm struck by how his experiences—both the hardships and the triumphs—have left an indelible mark on our family and everyone he meets. His story is a testament to the strength of the human spirit and the difference one can make when living by one's principles.

My father's journey from the battlefields of Vietnam to building a life rooted in learning and hard work is a powerful reminder of the potential within each of us. His legacy is not just in the tangible things he's built or fixed but in the lessons he's passed down—a legacy of courage, perseverance, and a commitment to always doing one's best.

CHAPTER ELEVEN

My Father's Silver Star

The job of a combat soldier is no joke. They would get exposed to enemy fire coming from any direction as they go out to fight the enemy. Not everybody has the courage to fight in a war. Truly, the gallantry that soldiers would deliver in battle is something worth being recognized. In the US Army, awarded to gutsy soldiers who have exhibited valor and heroism for a brief period like a time in a major battle is the Silver Star. Soldiers who have died in battle are also recognized with the Silver Star posthumously in care of the next of their kin. My father was awarded the Silver Star on November 4, 2017.

The Silver Star belongs to the military merits for valor. The military merits of valor honor soldiers who have done exceptional feats in battle and are the medals highest in rank among all the medals. The Silver Star is the third-highest in rank. It is the successor of the Citation Star, which denoted a citation for gallantry in action, established around the period of World War I on July 9, 1918, by the United States Congress. The Citation Star was a silver star measuring three-sixteenth of an inch or 4.8 mm. It would be attached to the service ribbon of the

World War Victory Medal when it was awarded to a soldier.

It was on July 19, 1932, when the conversion of the Citation Star into the Silver Star was approved by the United States Secretary of War. In the design of the Silver Star, the Citation Star was still incorporated at the center. The Silver Star is a gold medal shaped like a five-pointed star with a circumscribing diameter of one and a half inch or 38 mm. Surrounding the Citation Star in its center are rays that are encircled by a laurel wreath. At the back of the medal, inscribed are the words, "FOR GALLANTRY IN ACTION" along with the name of the recipient below. It is attached to a striped ribbon with the colors old glory red at the center and white and ultramarine blue at the sides.

The Silver Star came as a surprise to my father. One fateful day in 2017, my father just went to talk to their unit historian to ask him to make a change on the title on his name tag. Inscribed on his name tag was the title First Lieutenant. He wanted it replaced with First Lieutenant/Captain, so when he goes to military reunions, he would have First Lieutenant/Captain on it.

After he served as a forward observer in Vietnam with the rank of first lieutenant, when he had gotten out of Vietnam, he became a battery commander at Fort Sill and achieved the rank of captain. In his five years as a soldier in the army, he spent six months as a lieutenant serving as a forward observer in Vietnam, one and half years in

Walter Reed Hospital getting surgery after surgery and recovering from the injuries he got in Vietnam, and three years as captain with the role of a battery commander in Fort Sill.

To his surprise, the unit historian also informed him that he had been awarded the Silver Star. Because aside from asking for a change in the name tag, my father asked the unit historian if he could find out any information about a Bronze Star that he heard he had been recognized as eligible to receive in 1987.

The unit historian told him that there was no way to trace a Bronze Star after 50 years. On the bright side, the unit historian delivered to him even better news. "You should be happy you do have your four Purple Hearts and Silver Star," the unit historian said.

"What Silver Star?" My father asked in shock. The unit historian showed him a record wherein written was SS, some other abbreviations, subscript 2, (20 January 68). My father asked the unit historian what it meant. He explained to my father, "Six hundred people, including you, are on disc number two." My father had seen SS in his records but thought that the "SS" meant social security number. What it meant was Silver Star, and the date is written, "20 January February 68" was the date the Silver Star was awarded.

The unit historian also had a copy from the unit newspaper in Vietnam, where listed are men who went above and beyond their duty and had been awarded military merits such as Distinguished Service Cross, Silver Star, Bronze Star, Purple Hearts, etc. My father was one of four or five people awarded the Silver Star. My father could not believe it.

The historian said, "Well I have a record of it. I have proof." The unit historian mailed the Army record-keeping to ask for verification for my father's Silver Star and they verified that he, indeed, was awarded a Silver Star. My father, too, wrote to the Army record-keeping getting verification on his Silver Star.

When my father found out about his Silver Star 48 years later, he figured it may have been for the last battle he engaged in before he got out of the Army. That time when his recon sergeant, SP4Tim Nafe was among seven or so people who got killed and 35 to 40 people got wounded, which brought about casualties that were close to 75% of their unit.

During that battle, my father estimated that he may have fired around a million dollars' worth of ammunition. My father believed that he did a really tremendous job in that battle, so he thought it may have been for that. He was wrong. The Silver Star was for the courageous deeds he had done when their unit had their first encounter with the NVA in Dak To on November 7, 1967, that with his clever deduction, he helped give his unit an upper hand in combat and with his dedication, he continued to fight despite being wounded twice.

Looking back on that day, they were trapped by the

NVA. They were faced with the NVA from the top of Hill 724, and they had also been attacked with mortars being fired from behind them by NVA that took the position in the area they settled in the previous night. A mortar round coming from their nighttime position had wounded him.

Despite having been injured, he still carried out his job, directing fire towards the enemy with accuracy, that he managed to wipe out the NVA that attacked them from behind. He was also wounded for the second time by a grenade that went off near him, but he still fought with all the might he had left in him to subdue the NVA positioned at their front.

Nothing hindered him from doing his duty out there in the field, that even though he, himself, was a casualty, he still had the initiative to organize litter bearers and assist medics. Because of the serious injuries that he bore from that battle, he had to spend one month in Japan Army Hospital to recover until he was finally sent back to Vietnam on December 28, 1968.

When he received a letter back from the Army record-keeping, he perused it and finally got verification. He realized that the record of the recipients of the Silver Star that he was included in had been there all along, but never caught up to him.

Then he turned it over to Senator Pat Toomey. He was later officially awarded the Silver Star by Toomey on November 4, 2017.

Writing a book on his experiences in Vietnam soon

came to my father's mind. Some of his friends from Vietnam had also written books about the war. One of which was the battery commander of the 6th Battalion, 29th Field Artillery that his unit was assigned to.

There was also this other person, a captain, and lieutenant colonel, CPT/LTC Robert Barret had written about their unit in his book entitled *Central Highlands Redlegs*. So, he decided to have something written about what it was like to be out in the field.

He did not know at that time in the war, but he wiped out around 250 North Vietnamese soldiers that were in their nighttime position. None of the men he was with ever knew it either. They never went back to that hill to verify the number of NVA killed. It was not until approximately 50 years later when he went down to Florida to see a veteran who was dying that he got to talk to a radiotelephone operator (RTO) that he was able to have it verified.

The RTO said, "Yeah, they were firing from the hill that we just left." That finally gave him concrete proof that when he shifted the fire to their night-time position, he probably took out the NVA companies that came as reinforcements from behind with mortars, which he figured was a total of 250 NVA troops. He managed to wipe out that many NVA from behind, which was a good portion of a Vietnamese regiment.

He figured he got a couple of hundred people with the first round he fired. He also fired all sorts of ammo like 105s, 155s, 175s, and 8-inch guns and put out projectiles that weighed anything from 37 and a half pounds to maybe 200 pounds.

Valor Without Award: McNulty's January 19, 1968

In the 1970s, my father asked Robert Barrett—a fellow officer and friend—why so few men in their unit received medals for their service in Vietnam. Barrett replied simply, "In Vietnam, the focus was on mission success and survival—not medals."

That answer spoke volumes. Acts of valor were common, but formal recognition was rare. Soldiers were consumed with surviving, protecting one another, and enduring the physical and emotional toll of war. When they returned home, many carried their experiences in silence, choosing humility over honors.

My father never needed a medal to know he'd done his part. He believed every soldier had a role—infantry fought, medics healed, and forward observers directed artillery with precision. "We each did our part," he often said. "I was just doing my job."

In 2017, when Barrett began collecting stories for his book *The Central Highland Redlegs*, my father contributed a firsthand account of January 19, 1968—a day seared into his memory. Barrett already knew the broad outline, but the detailed account revealed the true magnitude of my father's actions—and the oversight that needed correcting.

After the book was published, Barrett assembled an awards packet recommending Lt. Pat McNulty for a second Silver Star. Just as he started gathering records, the COVID-19 pandemic shuttered the National Archives, delaying progress for years.

During that hiatus, Mark Coonrad—whose uncle served with the 3rd Battalion, 12th Infantry and was killed in action—created a Facebook group to reconnect veterans. Mark's archival research proved invaluable to the recommendation effort.

Those efforts confirmed what the men had long known: on January 19, 1968, my father wasn't coordinating fire from four firebases and a mortar unit, as previously believed. He directed artillery from seven firebases and two mortar teams—on foot, in mountainous terrain, under direct attack from a full North Vietnamese battalion.

Coordinating that many units under fire was staggering. Yet Lt. McNulty carried it out with composure and technical skill. His fire missions were so precise that subsequent patrols identified the enemy as a full NVA battalion. Stationed in the rear to better read the terrain, Lt. McNulty became the enemy's prime target. Despite the danger he remained focused, calling in life-saving fire adjustments. Of the 75 men in his unit, 68 made it back to base alive, though 27 were wounded.

One of the most compelling pieces of evidence came from the journal of then–Battalion Commander Colonel Hendrix, who observed the battle from a helicopter and came under fire himself. Hendrix trusted Lt. McNulty implicitly, entrusting him with all fire-control decisions. In a January 20 journal entry, he noted that McNulty had been wounded and was "in the thick of it the day before."

Using Hendrix's journal and the supporting records, Barrett submitted the awards packet through Senator Bob Casey to the Army Awards Committee. The committee requested a second officer's signature; Barrett explained he was the only surviving officer in McNulty's chain of command and provided documentation of the others' passing. The package was twice returned for lack of that signature.

In 2025, Barrett sent a follow-up letter asking the committee to accept Colonel Hendrix's journal as the required second signature—a firsthand account from a senior officer who witnessed the battle and trusted McNulty's leadership. The committee again declined, insisting on a second officers signature in order to proceed.

This effort isn't just about honoring one man. It's about setting the record straight—and recognizing the quiet leadership that saved lives without ever seeking recognition.

Unfortunately, Robert Barrett passed away shortly after this third request. My father suffered two strokes this past year and has recently been placed in hospice care.

CHAPTER TWELVE

My Father's Purple Hearts

Bloodshed and death are inevitable in war. It is awe-inspiring how the courageous soldiers would charge into battles, risking themselves with that high chance of getting wounded or even meeting their death for a particular objective. Indeed, those soldiers that have been wounded in action (WIA) and killed in action (KIA) deserve to be awarded for their sacrifice.

Since 1932, soldiers who have endured the excruciating pain of being WIA and have lost their lives fighting in war have been given recognition with the award of the Purple Heart.

My father, having been wounded four times while on active duty in the Vietnam War, had received four Purple Hearts. Highly revered and one of the oldest US military decorations, the Purple Heart is military merit that represents bravery in facing the enemy and the toll it takes on the troops. It originated from the Badge of Military Merit that was established by former president George Washington on August 7, 1782, when he was the commander-in-chief of the Continental Army. The Badge of Military Merit was awarded to soldiers who, according to the military order of the Badge of Military Merit, exemplified "not only instances of unusual gallantry in battle, but also extraordinary fidelity and essential service in any way."

It was a heart-shaped purple cloth with a silver braid on its edges that were to be sewn on the coat over the left breast. It was also Washington, himself, who awarded the Badge of Military Merit to soldiers that had been recognized during the Revolutionary War.

Only three soldiers had been awarded. It had not been granted again since. Several years later, on January 7, 1931, Gen. Douglas MacArthur confidentially pushed through work on the revival of that military merit, which was issued on the 200th anniversary of the birth of Washington on February 22, 1932. It came with a new design in the form of a medal instead of a piece of cloth. The design was sketched by an Army heraldic specialist in the Office of the Quartermaster General, Elizabeth Will according to general specifications.

The medal was then named the Purple Heart. The Purple Heart is a gold medal characterized by its heart shape detailed with a border encompassing a raised purple heart that has the profile relief bust of General George Washington in Continental Army uniform along with Washington's shield of the coat of arms in between green leaves at the top center of the heart. At the back of the medal is an embossed design of a heart shape with the inscription of "FOR MILITARY MERIT" along with the name of the recipient below.

The medal is attached to a purple ribbon with white edges. It is intended for commemorating meritorious service as well as recognizing soldiers that had been wounded and killed. The Military Order of the Purple Heart states that the Purple Heart is "Awarded to members of the armed forces of the United States who are wounded by an instrument of war at the hands of the enemy and posthumously to the next of kin in the name of those killed in action or who died of wounds received in action. It is specifically a combat decoration."

My father endured his first wounds two months into his military service in Vietnam. It was back on November 7, 1967, when they were headed to Hill 724 in Dak To. His first wound was when they were being attacked with a 51-caliber heavy machine gun and 60 mm mortars from an unknown location as they approached the hill. A mortar exploded near him, which hit his wrist and had even blown off his Timex watch. Nevertheless, he still managed to carry out his job as an FO.

On that same day, he got wounded for the second time by a US hand grenade, which he was unsure whether had gotten into the hands of the enemy or if it was friendly fire. It caused him to bleed internally in his lungs that he could not breathe and was spitting out blood. He thought he was going to die. Miraculously, he survived.

When he came back to Vietnam from hospitalization, he was on the move, still with a full cast in the arm that had been broken three places. The cast was later taken off in Vietnam.

The third time he was wounded was on January 19, 1968. The wounds he endured that day were the most serious ones he got in his whole military service in Vietnam. It happened right after their encounter with the enemy. They were expecting a counter-attack the night of their encounter. That night, they were awakened by three 82 mm mortars dropped at their base by the NVA. One exploded at an arm's length from him. Little did he know that it was that mortar that would put an end to his military service in Vietnam.

The mortar did a lot of damage to him including a couple of broken bones and a baseball-size chunk taken off in an arm, two nerves cut in his hand and fingers, an ulcerated eye, a hole in a femoral artery in one leg, a blown-out eardrum, and his entire body was covered in shrapnel.

It was when he was in the 71st Evacuation Hospital, being stabilized from his previous injuries that he had been wounded for the fourth time, as the hospital was attacked by the NVA with rockets and mortars. Recovery from his injuries was a long process. When he was finally stable enough to move after spending a few months in the 71st Evacuation Hospital in Vietnam for recovery, he then traveled to Japan and later was moved for further treatment in Walter Reed Army Hospital in Washington,

D.C. There, he had undergone several surgeries. It took one and a half years for him to fully recover.

During the six months that my father served as a forward observer in Vietnam, he said there had been little rough experiences in there, but the major events that had the most number of deaths were on November 7, 1967, and January 19, 1968.

The Vietnam War was not like the War on Terror (Operation Iraqi Freedom/Operation Enduring Freedom) where 4,000 people had been killed in 10 years. In the Vietnam War, the average KIA per week was 500.

CHAPTER THIRTEEN

The Quiet Hero: A Legacy of Sacrifice & Generations Saved

My father set out to be the best forward observer he could be. His friend, Sergeant Byron Kinnan, once said, "It was the forward observer (FO) who had to stay cool and face death head-on, to relay the correct coordinates, and to walk among those artillery explosions closest to the soldiers to protect their lives! To be precise and accurate, even while the battle raged around him, was the true test of any FO."

For an FO, it wasn't just about eliminating the enemy; it was equally about saving the lives of fellow soldiers.

Reflecting on his service, I realize that my father's heroism lay not only in engaging the enemy, but also in safeguarding the lives of his comrades. He not only saved lives on the battlefield but also preserved generations to come. Every soldier who returned home had the chance to start a family and have children and grandchildren.

If these men had perished in war, those future generations might never have existed. This underscores how profoundly each soldier's survival affects countless lives beyond their own.

There's no way to count the number of people my father, Pat McNulty, saved during the Vietnam War. On January 19th, 1968, during a fierce engagement, none of the 68 men in his unit would have made it back to the firebase without his quick thinking and the accurate directions he provided to the seven firebases and two motor units assisting with artillery support. Outnumbered and caught in a deadly trap, his skill in directing artillery fire was crucial to their survival.

After Vietnam, my father underwent 27 surgeries and was hospitalized for a year and a half. He had to adapt to living with only one eye and limited movement in his left arm, as part of it had been severely injured. He suffered from excruciating headaches for many years due to post-concussive syndrome caused by a mortar rounds exploding near him.

For over 50 years, he has endured nightmares that transport him back to the war, reliving those harrowing experiences as if they happened yesterday. Even today, pieces of shrapnel from the explosions continue to work their way out of his body—remnants of the past embedded deep within his flesh.

Despite all this, I have never heard him complain. I don't believe he would trade a single day of his time in Vietnam because he genuinely believes in God, Country, and Family.

My father holds himself to the highest standards in everything he does. He constantly asks himself, "Have I done everything to the best of my ability?" He has always pursued honor for his country and his family. His unwavering commitment and sacrifice have not only impacted those around him but have also ensured the continuance of many lives and families.

God bless all American soldiers for upholding the American way and for the generations of lives they have touched through their service and sacrifice.

Patrick John McNulty, Jr.: A Legacy of Valor, Perseverance & Dedication

Patrick John McNulty, Jr. was born on March 11, 1944. He is the oldest of eight children of Patrick and Alma McNulty. He graduated from LaSalle College in 1966 with a bachelor of science in Industrial Management degree. He attended ROTC during his four years at LaSalle and was commissioned as an Army 2nd lieutenant.

He then entered Active Duty. He also attended Ranger School and graduated from it in March 1967. In August of 1967, his first child was born 10 days before he landed in Vietnam to serve in the war as an artillery forward observer.

On November 7, 1967, he was severely wounded by a mortar round but continued to accurately direct fire toward the enemy. He was wounded a second time that day by a grenade and still kept going. Even though he was severely wounded himself, he organized litter bearers and assisted medics. His acts of valor on that day earned him the Silver Star.

He was sent to Japan Army Hospital, where he spent one month recovering from the injuries that had been inflicted by the mortar and grenade. He was then sent back to Vietnam on December 28, 1968. He was wounded for a third time by a mortar round. While he was being stabilized in the evacuation hospital, he was wounded for the fourth time, as the hospital was attacked by rockets and mortars. He spent one and a half years in Walter Reed hospital recovering from his injuries.

He was awarded four Purple Hearts for his injuries during active duty. Thereafter, he endured countless surgeries during his lengthy hospitalization. In June 1969, he returned to Fort Sill, Oklahoma, where he served as a battery commander and battalion adjutant. He retired as a captain.

Meanwhile, he married Alice Ann on August 27, 1966. They have been married for 54 years! They now have five children and ten grandchildren. Patrick attained his MBA from Temple University in 1974. He worked for Ford Motor Company as a senior industrial engineer. While working full-time for Ford, he built a house by himself. He did his own electrical work, plumbing, flooring, roofing, drywall – essentially everything was done by him. He is the textbook definition of a "do-it-yourselfer." As a world traveler, he loves to travel. He had been to Hawaii. He had been to Australia four times, Ireland 20 times, Germany and around Europe around 20 times or so. A favorite destination for his family when he would take them for family vacations would be Disney World, which they have been to over 25 times.

He is also known for his love of saving money and being thrifty. He has this mindset that "If there is no coupon for it, why buy it?" He keeps himself busy by maintaining his six properties and spending time refurbishing his historic house. Partick John McNulty's motto is: "The difficult we do today, the Impossible, takes a little longer."

He had also undergone Pistol Marksmanship Training, where he achieved the highest level, the distinguished expert for right-handed.

Impressively, he was able to do that even with disabilities in an eye and an arm. Moreover, no one had ever achieved that honor within 40 years in his gun club, which made him the first to earn it after four decades. He had also been working on attaining the distinguished expert for left-handed, and he is about 95% of the way through completion.

He is living proof that attitude is everything. He is a loving father and grandfather and we honor his examples of extreme courage, valor, and devotion to his family and country.

Afterword

This book, written on behalf of my father, is based entirely on a true story, with only three names changed. It was some time after he was officially awarded the Silver Star that the idea of sharing his experiences from Vietnam in a book came to mind. Everyone who went to war has his own story, and my father's is no exception. Although he considered writing it himself, he felt hesitant, as many veterans do when revisiting the realities of war. Nevertheless, I believe his story is one that deserves to be told, so I took on the honor of writing it for him.

My father carefully reviewed and edited this book after I completed the initial draft, ensuring his experiences were accurately and authentically represented.

Testimonials

"The first day I carried a radio for Sgt. Morfecies during a squad patrol, Capt. Bell sent out two patrols in different directions. The other patrol said they had heard movement and were going to call in a white phosphorus (WP) marking round. As we waited, I heard a "shot out" on my radio. It was a round, and it then took the treetops off over our heads way high up.

Sgt. Morfecies was Puerto Rican. I gave him the radio, and he was so excited, all he could come up with was Spanish and was speaking as fast as he could. I grabbed the horn and calmly explained that they were firing at us. Capt. Bell asked, "Who the hell was that? Is he on our side?" I never will forget that.

After that, I was very curious about how artillery was called in. Somewhere down the line, Pat McNulty explained to me exactly what was going on with artillery fire. I still remember him talking to me about correction and deflection and watching as he called in a protective wall of fire around us.

We were always surrounded by the enemy, however, and had incoming fire more often than we liked. We just humped the hills, set up defenses, and held our ground overnight. Capt. Bell and your Lt. Pat Mcnulty had artillery surround us as we moved from hill to hill."

—Glenn Sommers
Radio telephone operator (RTO) for Delta Company

"Back on November 7th on Hill 724, we were pinned down by a machine gun. Lt. McNulty was wounded and laying by a bunch of bamboos. I walked up to him, and he said, "Sarge, you have better get down before you are hit too. I responded with, which I shall never forget, "If you are on the ground, they will hit you in the head because they are in holes and small and shooting downhill." The last time I saw him, we laughed about that."

—Sgt. William West of D Company

"It's been a long time since we were together in the Snake Pit. The finest of the finest of the Vietnam War who became amputees. We all had more courage than sense. I was among them. I'm glad you're telling your story."

—Max Cleland
Former US Army captain and former US senator

"It was a great privilege to have served alongside a soldier so conscientious about his very important job. He certainly provided even danger close artillery as necessary to assist the infantry in their fight for survival."

—Jerry Afford, Staff Sergeant E-6 with D Co 3rd Battalion, 8th Infantry, 4th Inf Div

"Fifty years after the war, he flew all the way to Australia just to see me. He remembered me in my prime. We shared stories and laughter—no longer officer and PFC, just two boys, now older men, still brothers in arms. When he visited, it was a good visit—he even found a Catholic church for Pat and Alice to attend while he was here."

—William Ferguson (Doc Fergy) Medic, MP

"Five decades ago, Captain Partick John McNulty was severely wounded during a firefight near the Vietnam, Cambodia, and Laos border. The courage Capt. McNulty displayed to not just stay in the fight but to also save the lives of his brothers in arms is a true example of American heroism. It was an honor to play a small role in recognizing and celebrating Capt. McNulty's lifetime of exemplary service to our nation."

—Senator Pat Toomey from Pennsylvania

Special Thanks

I would like to thank all the veterans that my father had the pleasure to serve with or know after Vietnam. Special thanks to:

Recon Sergeant SP4 Tim Nafe
Capt. Terry Bell Tim Wilson - Medic
Bob Nadolski Capt. Max Cleveland Glenn Sommers, Radio Telephone Operator
Capt. Taylor L19 Birddog Air Controller Larry Skoglund Forward Observer
Bill Ferguson, Medic
Capt. Falcone
Capt. Taylor
Capt. Robert Barrett - Captain of firebase #25
Sgt. Byron Kinnan Sgt. Greenwood
Dr. Lt. Colonel Major
Walter Gross
Sgt Jerry Afford
Mike Anderson, Radio Telephone Operator
Bob Babcock – 4th Infantry Division / Deeds Publishing

Thomas Cooper, Medic Bravo Company
Sam LaBruno
Willie Glenn Powell
Gene K. Boyer
George Marchese
Bruce Bridenbaugh
Jim Cool
Mark Coonrad – helped create Facebook group
3rd of the 12th Ed Smola
Charlie Fisher

Informal Definitions, Abbreviations, Acronyms & Technical Details

AO – Area of Operations: The specific geographic region where a military unit carries out its missions.

AWOL – Absent Without Leave: When a soldier is missing from their post without official permission.

B40: A shoulder-fired grenade launcher used by North Vietnamese forces. Similar to the RPG-2, it fires a 40mm explosive projectile capable of damaging armored vehicles.

CA – Combat Assault: A direct attack against enemy forces.

CIB – Combat Infantryman Badge: An award given to U.S. Army infantrymen and Special Forces soldiers who have engaged in active ground combat.

Charlie: A term used by U.S. forces to refer to the Viet Cong or North Vietnamese Army (NVA). It comes from the NATO phonetic alphabet for "VC" (Victor Charlie).

CO – Commanding Officer: The officer in charge of a military unit.

CO HQs – Company Headquarters: The main command center for a company-sized military unit.

Clicks: Slang for kilometers. One "click" equals one kilometer.

C-rations: Canned food is provided to soldiers when fresh food is not available. They are individual meals ready to eat in the field.

DA – Department of the Army: The U.S. federal agency responsible for land-based military operations.

DEFCONs – Defensive Fire Concentrations: Preplanned artillery fire aimed at specific defensive locations.

DX – Direct Exchange: A system for exchanging damaged equipment or clothing for new items.

FB – Firebase: A temporary military camp that provides artillery support to infantry units in the field.

FDC – Fire Direction Center: The unit responsible for calculating and directing artillery fire.

FFZ – Free Fire Zone: An area where anyone can be considered hostile, allowing military forces to fire without additional permission.

FO – Forward Observer: A soldier who directs artillery fire by observing the enemy and providing coordinates.

Gooks: An offensive and derogatory term historically used during the Vietnam War to refer to Vietnamese people. **Note:** This term is highly offensive and should not be used.

HE – High Explosive: A type of explosive material that detonates rapidly.

H&I fires – Harassment and Interdiction Fires: Artillery fire aimed at disrupting enemy activities and movements.

INF – Infantry: Soldiers trained to fight on foot.

IRA – Irish Republican Army: An organization that sought Irish independence from British rule.

IV – Irish Volunteers: A military group established in 1913 advocating for Irish self-governance.

KIA – Killed in Action: Refers to soldiers who have been killed during combat operations.

LP – Listening Post: A hidden position near enemy lines used for surveillance and early warning.

LZ – Landing Zone: An area designated for helicopters or aircraft to land.

MOS – Military Occupational Specialty: A specific job or duty classification for which a soldier is trained.

MRE – Meal, Ready-to-Eat: Pre-packaged food rations for soldiers in the field.

NCO – Non-Commissioned Officer: An enlisted member with leadership responsibilities, such as a sergeant.

NVA – North Vietnamese Army: The regular military forces of North Vietnam.

PFC – Private First Class: An enlisted rank in the U.S. Army above Private and below Specialist or Corporal.

PEB – Physical Evaluation Board: A board that assesses a service member's fitness for duty due to injuries or disabilities.

PJ – Parachute Jumper: A soldier trained in parachute operations, also known as a paratrooper.

R&R – Rest and Recuperation: Leave time given to soldiers to rest away from combat zones.

REMF – Rear Echelon Military Forces: Slang for support troops stationed away from front-line combat.

ROTC – Reserve Officer Training Corps: A college program that trains students to become military officers.

RTO – Radio Telephone Operator: A soldier who operates a radio for communication.

RVN – Republic of Vietnam: The official name of South Vietnam from 1955 to 1975.

Tet Offensive: A major series of coordinated attacks by North Vietnamese forces during the Vietnamese New Year (Tet) in 1968.

Trophy: Slang for enemy equipment captured during combat.

UDT – Underwater Demolition Teams: Specialized units trained in underwater demolition and reconnaissance; precursors to the Navy SEALs.

USO – United Service Organizations: A nonprofit organization providing support and entertainment to U.S. military personnel.

UVF – Ulster Volunteer Force: A paramilitary group in Northern Ireland established in 1912.

Viet Cong (VC): A guerrilla force that fought against South Vietnam and U.S. forces, allied with North Vietnam.

Walking Point: The lead soldier or scout at the front of a patrol moving through potentially hostile territory.

WIA – Wounded in Action: Refers to soldiers injured during combat operations.

XO – Executive Officer: The second-in-command officer in a military unit.

.51-caliber heavy machine gun: A heavy machine gun used by enemy forces, firing large-caliber rounds.

60mm mortar: A lightweight mortar used for indirect fire support with a range of about 1,800 meters (1.1 miles).

81mm mortar: A medium mortar used by U.S. forces with a range of about 3,200 meters (2 miles).

82mm mortar: A medium mortar used by North Vietnamese forces with a range similar to the 81mm mortar.

105mm Howitzer: A towed artillery piece used by U.S. forces with a range of about 11,000 meters (6.8 miles).

155mm High Explosive (HE): A larger artillery piece capable of firing shells up to 14,600 meters (9 miles).

175mm gun: A heavy, self-propelled artillery piece with a maximum range of 33,000 meters (20.5 miles).

8-inch (203 mm) M110 self-propelled howitzer: A large artillery system with a range of up to 30,000 meters (18.6 miles) when using rocket-assisted projectiles

About the Author

Partick John McNulty, III, born in 1967 and raised in Sellersville, Pennsylvania United States, is the eldest son of Lt./Cpt Partick John McNulty, Jr, the Vietnam War veteran whose story he had written about in *In the Pursuit of Honor: A Vietnam War Veteran's Story*. Proud of his father's accomplishments not only as a service member of the US military but also in life, he wanted to share his father's story through writing a biography as a legacy book for his father. McNulty, III felt that he wanted to tell of Lt./Capt. McNulty committed service on active duty in the US military and how a great person he is through a book. McNulty, III is a psychiatric nurse and a professional photographer. As of now, he resides in Malvern, Pennsylvania with his wife Debbie and two daughters Michelle and Elizabeth.

Made in United States
Troutdale, OR
11/03/2026

56844873R00116